Edexcel

Functional Skills
Mathematics

Student Book
Level 1

Series Editor: Tony Cushen

Authors:
Chris Baston
Tony Cushen
Joan Knott
Alistair Macpherson
Su Nicholson
Carol Roberts

A PEARSON COMPANY

Published by Pearson Education Limited, a company incorporated in England and Wales, having its registered office at Edinburgh Gate, Harlow, Essex, CM20 2JE. Registered company number: 872828

Edexcel is a registered trademark of Edexcel Limited

The rights of Chris Baston, Tony Cushen, Joan Knott, Alistair Macpherson, Su Nicholson and Carol Roberts to be identified as the authors of this Work have been asserted by them in accordance with the Copyright, Designs and Patent Act, 1988.

First published 2010

13 12 11 10
10 9 8 7 6 5 4 3 2 1

British Library Cataloguing in Publication Data
A catalogue record for this book is available from the British Library

ISBN 978 1 84690 769 2

Typeset by Techset, Gateshead
Picture research by Katharine Oakes and Alison Prior
Printed in Great Britain at Scotprint, Haddington

Acknowledgements

Picture Credits

The publisher would like to thank the following for their kind permission to reproduce their photographs:

(Key: b-bottom; c-centre; l-left; r-right; t-top)

Alamy Images: vario images GmbH & Co.KG 51/2, John Cooper 104, Leila Cutler 21, Paul Doyle 40, GeoPic 113, Brian Harris 29, Tim Hill 34, Blend Images 15/3, Sinibomb Images 36/2, itanistock 15/2, Hideo Kurihara 59/2, Leslie Garland Picture Library 49/2, Motoring Picture Library 100/3, Barry Mason 59, MIXA 21/2, Nordicphotos 15, numb 96, Martin Thomas Photography 76, Realimage 19, Nick Scott 18, Charles Polidano / Touch The Skies 4, 22/2, Image Source 103, BlueMoon Stock 35, Andrew Twort 26, Steve Welsh 53, Gari Wyn Williams 36; **E M Clements Photography:** 37; **Getty Images:** Eightfish / The Image Bank 81, Ben Stansall 20/2, Dr. Marli Miller / Visuals Unlimited 80-81; **iStockphoto:** 49, Karen Arehart 110, Marina Bartel 112-113, William Berry 115, Jill Chen 24-25, Joe Clemson 95, katarina drpic 105, Julie de Leseleuc 54-55, NuStock 105/2, David Partington 50, Marcos Paternoster 109, photoGartner 100, Jason Stitt, 75, 114, 72-73; **Pearson Education Ltd:** Steve Benbow 100/2, BananaStock. Imagestate 52, Photodisc. Photolink 66, Digital Vision 14; **Rex Features:** Nigel R. Barklie 61; **Shutterstock:** 33, 48-49, 102, Vadim Balantsev 64-65, Philip Date 10-11/2, Jiri Hera 24-25/2, Blaz Kure 64-65/2, karam Miri 10-11, James Steidl 94-95, Christopher Walker 48-49/2; **Thinkstock:** 8, 106, Comstock 65, 65/1, 75, Comstock 65, 65/1, 75, iStockphoto 13, 16, 20, 26/2, 74, 85, iStockphoto 13, 16, 20, 26/2, 74, 85, iStockphoto 13, 16, 20, 26/2, 74, 85, iStockphoto 13, 16, 20, 26/2, 74, 85, Photodisc 25, Stockbyte 22, 51

Cover images: *Front:* **Shutterstock:** Norma Cornes

All other images © Pearson Education

Every effort has been made to trace the copyright holders and we apologise in advance for any unintentional omissions. We would be pleased to insert the appropriate acknowledgement in any subsequent edition of this publication.

We are grateful to the following for permission to reproduce copyright material:

Figures
Map on page 80 from Modern Day Landranger (TM) map, reproduced by permission of Ordnance Survey on behalf of HMSO, © Crown copyright 2010. All rights reserved. Ordnance Survey Licence number 100030901;

Tables
Table on page 100 adapted from http://www.ukma.org.uk/campaign/distanceinfo.aspx, © Copyright UK Metric Association; Table on page 115 adapted from http://maps.police.uk/view/?q=manchester&url, with permission of Greater Manchester Police; Table on page 116 adapted from http://www.sheffieldweather.co.uk/jul08vp2.pdf; Table on page 116 adapted from http://www.iammotoringfacts.co.uk/section6_3.html;

In some instances we have been unable to trace the owners of copyright material, and we would appreciate any information that would enable us to do so.

Every effort has been made to trace the copyright holders and we apologise in advance for any unintentional omissions. We would be pleased to insert the appropriate acknowledgement in any subsequent edition of this publication.

Disclaimer

This material has been published on behalf of Edexcel and offers high-quality support for the delivery of Edexcel qualifications. This does not mean that the material is essential to achieve any Edexcel qualification, nor does it mean that it is the only suitable material available to support any Edexcel qualification. Edexcel material will not be used verbatim in setting any Edexcel examination or assessment. Any resource lists produced by Edexcel shall include this and other appropriate resources.

Copies of official specifications for all Edexcel qualifications may be found on the Edexcel website: www.edexcel.com

Contents

Introduction

About Functional Skills Mathematics

Congratulations! If you are studying this book you are working towards the Edexcel Functional Skills qualification in Mathematics at Level 1.

Functional Skills are designed to give you the skills you need to be confident, effective and independent in education, work and everyday life.

In Functional Skills Mathematics, you will work on questions set in real life that will help you develop the skills you will need in life.

25 The RAF is using some Hercules Aircraft to take aid into a disaster area.
Each Hercules Aircraft can carry 20 000 kg of aid.

The table shows the aid the RAF has to take.

Item	Weight (kg) each pack	Number of packs
Medical kit	120	120
Tent	140	245
Blanket	94	450
Food	150	350

Think First!
Calculate the total weight.

Q How many Hercules Aircraft does the RAF need for the aid?

You will learn how to apply the mathematics you know to solve real-life problems and you will develop these important process skills:

Process skill	What it is	What it means	% of marks in the exam
Representing	Selecting the mathematics and information to model a situation	You decide how to tackle the problem you have been given	30–40%
Analysing	Processing and using mathematics	You apply your mathematics skills and understanding to solve the problem	30–40%
Interpreting	Interpreting and communicating the results of the analysis	You interpret the problem and make conclusions, justifying your response	30–40%

About this book

Edexcel Functional Skills Mathematics is specially written to help you achieve the Edexcel Functional Skills qualification in Mathematics at Level 1.

Eleven chapters based on familiar maths topics such as number and measures. Enables you to relate Functional Skills mathematics work to your GCSE work

Contents

examzone pages show you sample student answers with tips from the examiner, and reports on real exam questions

Apply what you have learnt by trying the Assessment practice questions that come after each group of chapters

The final Exam-style practice section helps prepare you for real exam questions

Check your answers to the shorter 'Let's get started' questions to see that you are on the right track

The features of the chapters

Each chapter begins with a KnowZone section:

Recap the key maths knowledge and understanding on the topic

Clear objectives tell you what you will learn in the section

Worked examples show you how to solve functional-style questions

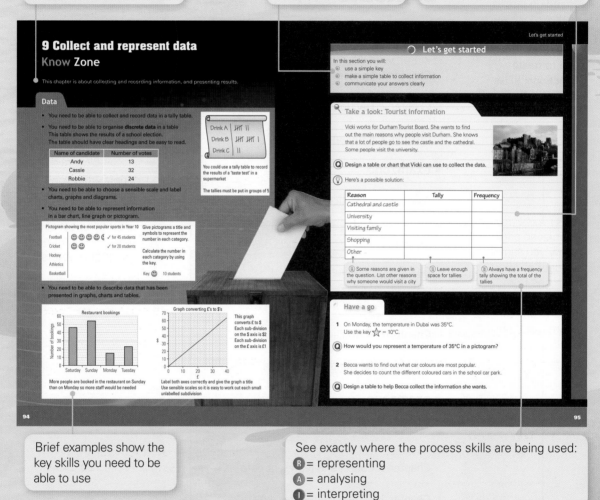

Brief examples show the key skills you need to be able to use

See exactly where the process skills are being used:
R = representing
A = analysing
I = interpreting

After KnowZone, three sections provide progressively more challenging questions for you to practise:

Let's get started short, straightforward questions to get you started

We're on the way more challenging questions with some of the features of real functional maths questions

Exam ready! longer and more complex questions to help prepare you for the exam

1   This spinner is used in a game.

Q Which colour is the spinner **least likely** to land on? Why?

2   Last week Gavin had an accident and broke his arm.

Q What is the likelihood of him breaking his other arm this week?

3   On his way to work today, Wes found £10 in the street.

Q What is the likelihood that he will find £10 in the street tomorrow?

> Plenty of practice questions for you to try

Q Farhan does not have all the information he needs to decide which age group is the safest.
What other information does he need?

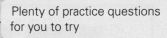

ResultsPlus
Exam Tip

Set out your working clearly, so that the examiner can see how you decided on your answer. Give reasons for your decision.

> This gives exam advice, useful hints and warns you of common mistakes that examiners frequently see students make

He does a survey.

Q List the possible methods of travelling to college.
Use your experience to describe the likelihood of each method of travelling.

7   A couple want to have three children.
It is equally likely that each child will be a boy or a girl.

Think First!

Think about what order the children are born in.

Q List the possible outcomes for the three children.

> Use the Think first boxes for tips to help you get started on a question

Q Use the chart to work out how much Stephanie's daughter has grown between her fourth and fifth birthdays.

4   Donna is feeling ill. The thermometer shows her temperature.

34   35   36   37   38   39   40

Fact

The human body's normal temperature is 36.8°C.

Q Donna's temperature is above normal body temperature.
How many degrees above?

5   Karen has been on a diet. She has lost 3.5 kg.

> Facts, abbreviations and specialist terms explained

Now you can:

Decide if events are certain, impossible, likely or unlikely
Find all the possible outcomes of an event
Know that some events can happen in more than one way
Know when events are equally likely
Use real data to decide on the likelihood of events

> Check your progress using this summary box at the end of the chapter

About ActiveTeach

Use ActiveTeach to view the course on screen with exciting interactive content.

Use our new ResultsPlus Problem Solving tool to plan and solve functional skills style questions

ResultsPlus Knowledge Check – Check your maths skills before you start each chapter

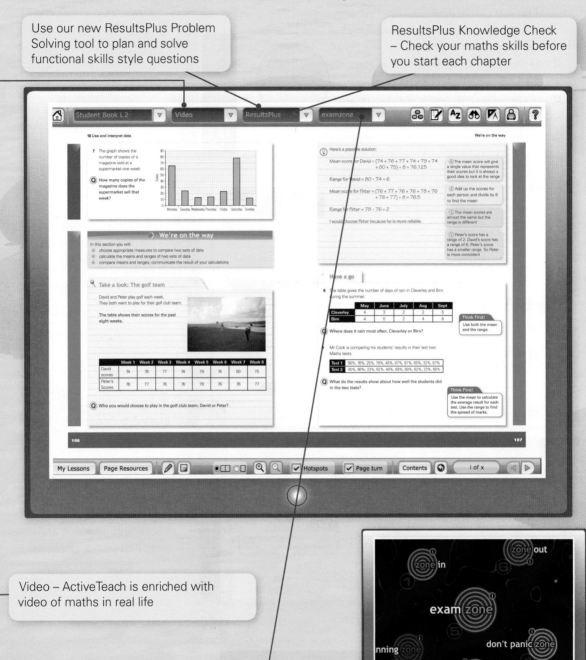

Video – ActiveTeach is enriched with video of maths in real life

examzone – provides a range of exam preparation resources including exam style practice with marked student answers and examiner feedback

The Functional Skills Mathematics assessment

You will take one paper which will be marked by Edexcel. The exam is 'pass' or 'fail' – there are no other grades.

There are many opportunities a year to sit the exam and you can take the exam as many times as you and your centre wish.

The exam lasts 1 hour 30 minutes and you can use a calculator throughout.

The exam paper has three sections. Each section has a theme, such as jobs or the garden.

Section 1 – first theme	16 marks
Section 2 – second theme	16 marks
Section 3 – third theme	16 marks
Total	**48 marks altogether**

You will gain marks for showing your working. Writing just the answer may not be enough to get full marks. So practise showing your working at all times.

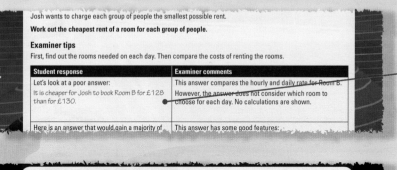

The examzone pages show you examples of poor student answers where working is not clearly displayed

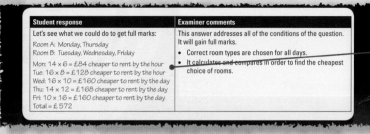

This student has shown all their working clearly and would gain full marks

1 Number
Know Zone

Numbers are very important in many aspects of our lives. Learning to count is a skill we learn when we are very young. In this chapter you will learn to apply that skill in many different situations. Some of the situations will be new to you.

Whole numbers

- You need to understand and be able to use whole numbers

Example Old Trafford's new all-seater stadium has a capacity of about seventy-six thousand.

> Capacity is the number of people that the stadium holds

Write seventy-six thousand in figures.

M millions	HTh hundred thousands	TTh ten thousands	Th thousands	H hundreds	T tens	U units
		7	6	0	0	0

Answer The ground holds 76 000 people.

Negative numbers

- You need to recognise and be able to use negative numbers in practical contexts

Example The temperature in Moscow was −8°C. The temperature in Berlin was −4°C.

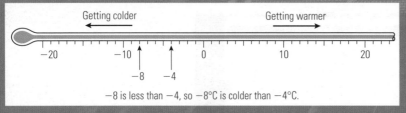

Getting colder ← | → Getting warmer

−20   −10   −8 −4   0   10   20

−8 is less than −4, so −8°C is colder than −4°C.

Which city had the lower temperature?
Answer Moscow had the lower temperature.

Add, subtract, multiply and divide

- You need to understand how to add, subtract, multiply and divide whole numbers using a range of methods

Example There are 13 cans of beans on a shop shelf. There are 24 cans of beans in the shop storeroom.
How many cans of beans does the shop have altogether?
Answer $13 + 24 = 37$ cans altogether.

> 'Altogether' is often a clue that you need to do an addition.

Example There are 48 bags of peanuts in a carton.
A shop sells 39 of the bags. How many bags of peanuts does the shop have left?
Answer $48 - 39 = 9$ bags left.

> 'Left' is often a clue that you need to do a subtraction.

Example A shop buys 5 cartons of bags of crisps to sell.
There are 48 bags in each carton of crisps.
How many bags of crisps are there in 5 cartons?
Answer $48 \times 5 = 240$ bags.

Example A shop puts 72 cans of soup into 6 piles.
How many cans will there be in each pile?
Answer $72 \div 6 = 12$ cans.

Multiply and divide whole numbers by 10 and 100

- You need to understand how to multiply and divide whole numbers by 10 and 100 using mental arithmetic

Example A shop sells boxes of eggs.
There are 15 eggs in each box.
How many eggs are there in 100 boxes?
Answer $15 \times 100 = 1500$ eggs.

Example A company has to pack 6543 model soldiers into boxes. The company can pack 10 models into each box.
How many boxes does the company need?
Answer $6543 \div 10 = 654.3$
The company needs 654 boxes for 6540 model soldiers.
It will need 1 more box for the 3 model soldiers left.
The company will need 655 boxes altogether.

◉ Let's get started

In this section you will:

- ◉ write numbers using words and figures
- ◉ choose the number operations you use to solve problems
- ◉ use place value to order numbers
- ◉ multiply or divide by 10 or 100
- ◉ interpret scales using negative numbers

🔍 Take a look: Writing a cheque

Here is a cheque to pay a bill of £5030
to Furniture Direct.
There is a mistake on the cheque.

➤ **DB Bank**   16-12-79
DATE _11ᵗʰ July 2010_
Pay to   FURNITURE DIRECT
Five hundred and three thousand   £ | 5030
pounds only
Signature _W. Brown_
".018373" 05:63594: 149573"

Q **Explain what is wrong.**

💡 Here's a possible solution:

The amount when written in words should say five thousand and thirty pounds.

🔍 Take a look: Ordering cheque numbers

➤ **DB Bank**   16-12-79
DATE _____
Pay to
£ [   ]
Signature _____
".018373" 05:63594: 149573"

A bank gets cheques on accounts with the following account numbers:
100467   298721   312452   190234   234765   100432   219837

❶ The digit at the left has the highest value

Q **Write down the account numbers in order.**
Start with the lowest account number.

💡 Here's the solution:

100432   100467   190234   219837   234765   298721   312452

Take a look: Booking a holiday

Hotel Sunscape
*Traveller Rating ****

4 days	Room	Holiday Type	Cost
2 Adults	Twin beds	half board	£335
2 Adults	Twin beds	full board	£440

Mr and Mrs Jones are going to have a 4-day holiday at the Hotel Sunscape.

Q What is the difference in cost between a half board and a full board holiday?

Here's a possible solution:

Cost of full board = £440
Cost of half board = £335
Difference = £440 – £335 = £105
Full board costs £105 more than half board.

 Have a go

1 Joe has to build a wall that has 10 rows of bricks.
There will be 142 bricks in each row.

Q How many bricks does Joe need?

2 Jill needs to pay for her holiday.
The holiday costs £4025

Q Write £4025 in words.

3 Manchester City Football Club has hospitality seats which cost £75 per seat.
It has 1500 hospitality seats.

Q How much money will Manchester City get when all the seats at a football match are full?

4 The following table shows the capacity of some football stadiums.

Team	Stadium	Capacity
Chelsea	Stamford Bridge	42 449
Liverpool	Anfield	45 362
Manchester City	City of Manchester	48 000
Hull	Kingston Communications	25 504
Wolverhampton	Molineux	29 400
Everton	Goodison Park	40 569

> **Fact**
>
> Capacity means the number of people a stadium can hold.

Some of these football stadiums are going to be used for international football matches. There will be about 30 000 people at each match.

Q **a)** List the stadiums in order of size.

b) Suggest three stadiums to use.
Give your reasons.

> **Think First!**
>
> Think of a reason why the biggest stadium may not be the best stadium to use.

5 Old Trafford football ground has 75 957 seats.
Wembley Stadium is the biggest stadium in the country. It has 90 000 seats.

Q How many more seats does Wembley have than Old Trafford?

6 Heights above sea level are described with positive numbers in metres.
Depths below sea level are described with negative numbers in metres.

Q What do the following mean in terms of sea level?

a) +5 m **b)** −10 m

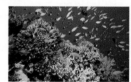

7 The table shows the temperature one night for five cities in the British Isles.

	Midnight
Manchester	−3°C
Newcastle	−2°C
Glasgow	−4°C
London	−1°C
Southampton	1°C

> **Think First!**
>
> Negative numbers are smaller than positive numbers.

Q Which of the cities were warmer than Newcastle?

8 A factory makes biscuits. A worker weighs 20 packets of the biscuits each hour to check that the packets aren't too far over or too far under 270 g.
The worker records packets of 280 g as +10.
The worker records packets of 260 g as −10.

Q How would the worker record the following weights?

a) 290 g **b)** 250 g **c)** 265 g **d)** 275 g

9 Alex needs 42 m of border for a room.
He buys 37 m of border. He did the calculation 37 − 42 on his calculator.
The calculator shows −5.

Q Explain what −5 means for Alex's situation.

Think First!
You must explain in terms of the length of the border.

10 Carole earns £1350 per month.

Q What is her annual salary?

Think First!
How many months are there in one year?

11 Jean is organising interviews for office assistants.
She is going to interview 34 people.
She can interview 13 people each day.

Q How many days does she need for the interviews?

12 A coach has seats for 28 people. 135 people are going on a trip.

Q **a)** How many coaches will be needed?
b) How many spare seats will there be?

Think First!
You can't leave anyone behind so consider which way you need to round.

⊙ We're on the way

In this section you will:
- choose the information you need to use
- use appropriate mathematical operations
- interpret results and solutions in the context of the question
- make conclusions

🔍 Take a look: Mallory Towers

Mallory Towers
House and Gardens
Adults £7
Children £3
Family (2 adults & 2 children) £17

Mr and Mrs Bakir want to take their 2 children and
their children's friends to visit Mallory Towers.
Mr and Mrs Bakir have a budget of £30.

Q How many of the children's friends can they take?

Here's a possible solution:

The cost of two adult tickets = 2 × 7 = £14.
The cost of two children's tickets = 2 × 3 = £6.
The cost of buying individual tickets for two adults
and two children = £14 + £6 = £20.

> **R** **A** Strategy: first find out what it costs to take the family!

They should buy a family ticket at £17.
This leaves £13 for them to spend on tickets for
their children's friends.

> **R** How many children's tickets can they buy for £13?

Number of tickets is
13 ÷ 3 = 4.33
They can take 4 friends.

> **I** 4 × £3 = £12. There is £1 left over in the budget

Take a look: Concrete

A builder needs to buy some materials. He has a budget of £500.

He needs to buy 2 cubic metres of concrete for a floor.
He wants to spend the rest of his budget on paving slabs for
paths around the house.

Paving slabs cost £85 per square metre.
Concrete flooring costs £95 per cubic metre.

Here is the order form for Concrete Plus.

 Concrete Plus
Concrete for all your needs

	Quantity	Cost
Paving slabs @ £85 per square metre	m²	£
Concrete flooring @ £95 per cubic metre	m³	£
	Total	

> **Fact**
> m² = square metre
> m³ = cubic metre

Q Complete the order form.

 Here's a possible solution:

2 cubic metres of concrete flooring costs 2 × £95 = £190
This leaves £500 – £190 = £310
2 square metres of paving slabs costs 2 × £85 = £170
3 square metres of paving slabs costs 3 × £85 = £255
4 square metres of paving slabs costs 4 × £85 = £340

> **R** The builder has £310 to spend on paving slabs. How many square metres of paving slabs can he buy for £310?

He can buy 3 square metres.

Four square metres will cost too much.

Concrete Plus
Concrete for all your needs

	Quantity		Cost
Paving slabs @ £85 per square metre	3	m²	£255
Concrete flooring @ £95 per cubic metre	2	m³	£190
		Total	£445

ℹ Check that the total cost is within the £500 budget set

⊕ Have a go

13 A pack of 50 bricks costs £15. A pack of 500 bricks costs £110. A bricklayer needs 13 650 bricks.

Q Work out the cost of the bricks.

Think First!

First work out how many of each pack you will need. The builder will want the lowest cost!

14 John counted the numbers of boys and the number of girls born in a hospital one weekend.

23 boys and 11 girls were born.

John says, 'There were twice as many girls as boys born over the weekend.'

John is wrong.

Q What should John have said?

15 The Direct.Gov website tells elderly people in 2010:

This year we will give you £25 when the temperature is 0°C or less for seven consecutive days.

Think First!

On how many days was the temperature below 0?

The calendar shows the lowest temperatures recorded in one town for each day in December.

Monday	Tuesday	Wednesday	Thursday	Friday	Saturday	Sunday
	1 1°C	**2** −3°C	**3** −5°C	**4** −2°C	**5** 1°C	**6** 2°C
7 2°C	**8** −1°C	**9** −3°C	**10** −3°C	**11** −2°C	**12** −1°C	**13** −4°C
14 −2°C	**15** 1°C	**16** 2°C	**17** 4°C	**18** 3°C	**19** 2°C	**20** 1°C
21 3°C	**22** 3°C	**23** 2°C	**24** 1°C	**25** −2°C	**26** −3°C	**27** −1°C
28 −1°C	**29** −3°C	**30** −6°C	**31** −6°C			

Q Will elderly people in the town get £25?

16 Joe has a small field.

50 m

10 m

He wants to put a fence around the field.
Joe needs to put a fence post in each corner of the field.
He needs to put other fence posts no more than 3 m apart.

Q How many posts does Joe need?

17 Sue is marketing manager for a company.
She has a budget of £60 000.
She is going to advertise in a newspaper.

The table below shows how much the paper charges for adverts.

Time period	Monday–Thursday	Friday or Saturday	Sunday
Full Page	£6400	£17 000	£18 000
Half Page	£3600	£9500	£10 500

Sue must advertise in every time period at least once.
She wants to spend as much of her budget as possible.

Q Advise Sue how she should spend her budget.

18 A football club is organising coaches for 857 supporters
to travel to put an away match.
A coach firm has the following types of coaches:

Type
15 Seat Minibus
46 Seat Luxury Coach
75 Seat Double Decker

Think First!

Try to fill all of the
seats using the
smallest possible
number of coaches.

Q Write down the coaches the
club should hire.

Exam ready!

In this section you will:
- choose the mathematical information needed to solve a problem
- use appropriate mathematical procedures
- find and interpret results and solutions
- make conclusions
- present and communicate results

Take a look: Gas bills

DEF gas company charges 8p for the first 617 units of gas.
It charges 3p for each extra unit of gas.

ONE gas company charges 7p for the first 1266 units of gas.
It charges 2p for each extra unit of gas.

A family uses 1900 unts of gas.

(Q) Which gas company is best for the family to use?

(☺) Here's a possible solution:

Using DEF:
617 units cost 617 × 8p = 4936p
1900 – 617 = 1283 units
1283 units x 3p = 3849p
Cost of 1900 units is 4936 + 3849 = 8785p
DEF costs £87.85

> **R** Compare the price of 1900 units with DEF and ONE Energy.

Using ONE:
1266 units cost 1266 × 7p = 8862p
1900 – 1266 = 634
634 units x 2p = 1268
1900 units cost 8862 + 1268 = 10130p
ONE Energy costs £101.30

So DEF is cheaper for 1900 units. ●————————

> **I** Don't forget to give your decisions

Take a look: Eggs

A farm shop shop sells eggs in two sizes of egg boxes.
There are 10 eggs in a large box.
There are 6 eggs in a small box.
The shop has 348 eggs.

(Q) How many of each size of box does the shop need for the eggs?
Give a reason for your choice of boxes.

(💡) Here's a possible solution:

(R) r = remainder

348 ÷ 10 = 34 r 8   34 large boxes, 8 eggs left over.
8 ÷ 6 = 1 r 2   1 small box, 2 eggs left over.
Use 34 large boxes and 1 small box. 2 eggs are left over.

Fact

There are many different ways of putting all the eggs in these two box sizes.

(💡) Here's another possible solution:

300 ÷ 10 = 30   30 large boxes
48 ÷ 6 = 8   8 small boxes
Use 30 large boxes and 8 small boxes.

(ℹ) This solution puts all of the eggs in boxes

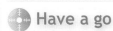 **Have a go**

19 Before 1990, if all the seats were full, many extra people stood up to watch a football match.
In 1990 Health and Safety regulations said that everyone at a football match must have a seat.
The table below shows information about the number of people attending football matches.

Team	Stadium	Largest ever capacity	Stadium capacity in 2010
Aston Villa	Villa Park	76 588	43 000
Fulham	Craven Cottage	49 335	22 000
Stoke City	Britannia Stadium	28 218	28 218
Tottenham Hotspur	White Hart Lane	73 038	36 214
West Ham	Boleyn Ground	42 332	35 647
Wigan Athletic	DW Stadium	23 575	25 000

(Q) **a)** Work out the difference between the largest ever capacity and the capacity in 2010 for White Hart Lane.
b) Which stadiums were built before 1990?

20 The table below shows the number of vegetables that Josie wants to plant.

Vegetable	Carrots	Cabbages	Peas	Beans
Number of plants	43	24	26	19

Josie has three vegetable plots (A, B and C). She can plant 40 vegetable plants in each plot. Plot A is **not** suitable for peas or beans. Plot B should **not** be used for cabbages or carrots. Josie can plant different vegetables together in the same plot.

Q Decide how many of each vegetable Josie can plant in each of the plots.

21 Amy is looking for a job. She finds two similar jobs advertised in her local newspaper.

> **Mega Store**
> *Sales Assistant*
> Pay £270 per week
> (40 hour week)

> **Supa Store**
> *Shop Assistant*
> Pay £8 per hour
> 38 hour week

Compare the pay for the two jobs.

Q Which job is better paid per week?

> **Think First!**
> You need to remember that the number of hours per week is different.

22 People can have four different types of treatment at the Beauty Box Salon.
They can have facials, eyes, spray tan and nails.
The prices are shown below.

The Beauty Box Salon			
Facials		**Spray tan**	
Aromatherapy	£40	1 session	£12
Luxury	£50	2 sessions	£18
		3 sessions	£24
Eyes			
Eyebrow shape	£10	**Nails**	
Eyebrow tint	£8	Manicure	£22
Eyelash tint	£12	Luxury manicure	£26
Eyelash and eyebrow tint	£15	Half hour manicure	£15

Sheila has a voucher for £100. She would like to have at least one of each type of treatment. She must not spend more than £100.
Sheila wants to spend as much of the £100 as possible.

Q Suggest four treatments that Sheila can have.

23 The table below shows some breakfast foods and the number of calories in the food.

Food	Weight (grams)	Calories
Bread (1 slice)	35	76
Cornflakes	30	108
Muesli	60	218
Baked beans	115	100
Bacon (grilled)	45	152
Butter	15	110

Food	Quantity	Calories
Milk (semi-skimmed)	140 ml	70
Egg (poached or boiled)	1	88
Coffee (black)	200 ml	4
Tea	200 ml	0
Orange Juice	140 ml	55
Sausages (grilled)	1	160

John plans to have 1 slice of bread, 90 g of bacon, 2 eggs and 140 ml of orange juice for breakfast.

He wants to have less than 550 calories.

Sue is on a diet. She must have between 300 calories and 500 calories at breakfast.

Q **a)** Will John's breakfast be less than 550 calories?

b) Plan a breakfast for Sue.

24 This table gives a list of the calories burnt during 20 minutes' exercise.

Exercise	Calories
Walking	80
Cycling	160
Running	90
Aerobics	140
Cleaning	50
Driving	35
Gardening	160

One Saturday morning, Mai cycles for 10 minutes to get to a sports centre.
She does an aerobics class for 1 hour. She then cycles home again.
In the afternoon, Mai does two hours of gardening.

Q How many calories does Mai burn?

25 The RAF is using some Hercules Aircraft to take aid into a disaster area.
Each Hercules Aircraft can carry 20 000 kg of aid.

The table shows the aid the RAF has to take.

Item	Weight (kg) each pack	Number of packs
Medical kit	120	120
Tent	140	245
Blanket	94	450
Food	150	350

Q How many Hercules Aircraft does the RAF need for the aid?

Think First!

Calculate the total weight.

26 John wants to hire a car for five days.

He can hire the car from one of these two companies.

| Most Motors |
| £45 per day plus 25p per mile |

| Collect a Car |
| £300 per week |

Think First!

'Compare' means you need to use words like 'more', 'less', 'bigger', 'smaller' and 'the same' in referring to the costs.

(Q) Compare the cost of hiring a car from each company.

27 Scot Power electricity company charges 12p for the first 174 units of electricity.

It charges 11p for each extra unit of electricity.

AB Energy company charges electricity 15p for the first 227 units of electricity.

It charges 9p for each extra unit of electricity.

(Q) **a)** Work out the cost of buying 700 units of electricity from Scot Power.

Mike uses between 700 and 1000 units of electricity.

b) Which electricity company should Mike use?

Now you can:

- Write numbers using words and figures
- Use place value to order numbers
- Multiply or divide by 10 or 100
- Interpret scales using negative numbers
- Choose the mathematical information needed to solve a problem
- Use appropriate mathematical procedures
- Find and interpret results and solutions
- Make conclusions
- Present and communicate results

2 Fractions, decimals and percentages
Know Zone

You can use fractions, decimals and percentages in many different practical problems.

Fractions

You need to know how to:

- Read, write and compare common fractions

Example Which is smaller: $\frac{2}{5}$ or $\frac{1}{2}$? **Answer** $\frac{2}{5} = 0.4, \frac{1}{2} = 0.5$ so $\frac{2}{5}$ is smaller.

- Write fractions in their simplest form

Example $\frac{12}{36}$ in its simplest form is $\frac{1}{3}$, dividing both the **numerator** and the **denominator** by 12.

- Work out common equivalent fractions

Example $\frac{6}{24}$ and $\frac{3}{12}$ both simplify to $\frac{1}{4}$, so they are equivalent fractions.

> Common fractions are halves, thirds, quarters, fifths and tenths

- Find common fractions of quantities and measurements

Example To find two-fifths of 45, work out $\frac{2}{5} \times 45$.

> **Fact**
>
> The fraction key looks like this: $\boxed{a\frac{b}{c}}$ or $\boxed{\frac{\blacksquare}{\square}}$

- Use the fraction key on a calculator to write fractions in their lowest terms or to calculate with fractions

Example To simplify $\frac{16}{20}$, key in 16 20 $\boxed{=}$

The calculator displays $\boxed{4 \lrcorner 5}$. Write your answer as $\frac{4}{5}$.

Decimals

You need to know how to:

- Order and compare decimals

Example 3.2 is more than 3.04

> This is because the 2 in 3.2 represents 2 tenths; the 4 in 3.04 represents 4 hundredths

- Add or subtract decimals to 2 decimal places. Problems are often about money or measurements.

Example 5.2 metres + 1.03 metres = 6.23 metres.

Percentages

You need to know how to:

- Identify equivalent common fractions, decimals and percentages. Learn the decimal and percentage equivalences for tenths, fifths, quarters, thirds and halves.

Examples $0.2 = \frac{1}{5} = 20\%$, $0.25 = \frac{1}{4} = 25\%$.

- Find common percentages of quantities using mental methods

Example 10% of £4.60 = £4.60 × 0.1 = £0.46

So 20% of £4.60 = £0.46 × 2 = £0.92; 5% of £4.60 = £0.46 ÷ 2 = £0.23

Let's get started

In this section you will:
- find fractions of quantities
- add or subtract decimals up to 2 decimal places
- round decimals or amounts of money
- use common decimal, fraction and percentage equivalences to find solutions to practical problems

Take a look: Shopping

Here is part of the price list for a market stall.

Baking potatoes	£1.58 per kilo
Milk	98p per litre
Minced beef	£3.60 per kilo

Mary buys 2 kilos of baking potatoes,
2 litres of milk
$\frac{1}{2}$ a kilo of minced beef.

R Note the exact quantities Mary has bought

Q How much does Mary pay?

Here's a possible solution:

Cost of potatoes: 2 × £1.58 = £3.16
Cost of milk: £1.96
Cost of minced beef: $\frac{1}{2}$ × £3.60 = £1.80
Total cost: £6.92

A Convert the cost of milk from pence to pounds

Have a go

1 A student wants to work out $\frac{1}{16}$ of £82.
He keys in 82 ÷ 16 = on his calculator.
The display reads 5.125.

Q What answer should the student write down?

2 Iris paid £1.37 for half a kilo of grapes from a shop.

Her neighbour bought a kilo of grapes from the supermarket.She paid £2.40
The grapes were more expensive in the shop than in the supermarket.

(Q) How much more expensive?
Show all your working.

3 Mandy hires a car with a full tank of petrol.
The hire car company tells her to bring back the car with the tank
$\frac{1}{4}$ full of petrol.
Mandy uses nearly all the petrol in the tank.
She estimates that a full tank is about 45 litres of petrol.

(Q) Approximately how many litres of petrol should Mandy put into the tank?

4 John wants to buy a shirt.
The same make of shirt is sold in two different shops.

Jack's Clothes Store
Shirt
Normally costs £48
Now has 25% OFF

Brown's Clothes Store
Shirt
Normally costs £60
Now has $\frac{1}{3}$ OFF

(Q) Which shop should John choose?

5 Three friends want to go to an Italian restaurant.
They find the following adverts for two restaurants:

Gino's
Main courses normally
£10.25 each
Special offer – Two main
courses for the price of one

Bella Belissima!
Main courses normally
£11.50 each
Special offer – Get 50% off
the price of every main course

(Q) Which restaurant should the three friends choose?

We're on the way

In this section you will:
- use equivalent fractions, decimals and percentages to solve practical problems
- find common fractions or percentages of quantities to solve practical problems
- use your working to explain and interpret solutions

Take a look: Smoking habits

A company has had a 'Give up smoking' campaign for the past year.

Here is information about the number of male smokers in the company.

	Fraction of male employees who smoke	Total number of male employees
Last year	$\frac{1}{3}$	36
This year	$\frac{1}{5}$	35

STOP QUIT WHILE YOU'RE AHEAD

Approximately 20% of men in the UK smoke.

Q Compare the number of smokers last year and this year with the UK percentage.

Here's a possible solution:

Last year's figures: $\frac{1}{3}$ is roughly 33%
This year's figures: $\frac{1}{5}$ is 20%
Last year, there were about 13% more male smokers at the company than the UK percentage of male smokers.
This year the percentage of smokers at the company was the same as the UK percentage.

Have a go

6 Sara takes three different tests. Here are her results.

IT (out of 36 marks)	Maths (out of 40 marks)	Science
12	20	40%

Q Compare her results in the three tests.

7 Danny owns a clothes shop.

He buys 60 pairs of jeans to sell in his shop. Danny pays £8 for each pair of jeans.

He sells $\frac{1}{4}$ of the jeans for £16 each pair.

Danny then has a sale.

He sells all the jeans that are left for £10 each pair.

(Q) What is the total amount of money Danny made?
Show all your working and explain your reasons.

8 Bill is the manager of a leisure centre.
He keeps a record of the number of people coming to each class for one month.
Bill shows this information in a bar chart.

The maximum number of people for each class is 200 people per month.
If less than 25% of the maximum attendance of people come to a class, Bill will close the class.

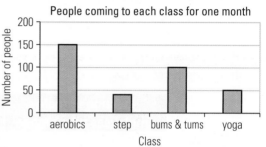

People coming to each class for one month

(Q) Which class will Bill close?
Show all your working.

9 Alan buys an old canal boat.

He wants to change how space is used inside the boat.

He draws this plan.

Bathroom and bedroom area	Kitchen and living area

← 20 m →

Alan wants the length of the bathroom and bedroom area to be about $\frac{2}{5}$ of the length of the boat.

(Q) What will the length of the kitchen and living area be?
Show all your working.

10 Joyce is going to sell cups of orange drink at a village fair.

She wants to sell 200 cups of the orange drink.

Each cup will hold 250 ml of the orange drink.

Joyce will mix 10% of orange squash with water for each cup of drink.

(Q) **a)** Work out how many litres of orange squash Joyce needs to buy.

Here are the prices of orange squash.

1 litre bottle	1.5 litre bottle
£1.50	£2

> **Fact**
> 1 litre = 1000 ml

b) What is the cheapest way to buy this amount of orange squash?

11 Danielle is thinking of applying for one of the following two jobs:

Telesales assistant
Salary: £13 000
per year

**Office clerk in a
city centre office**

Mon–Fri: 35 hours per week
Saturday: 5 hours
Rate of pay: £8 per hour
(Saturdays: time and a quarter)
Permanent position

Think First!

Work out a quarter of the hourly rate. Add this on to the hourly rate to find 'time and a quarter'.

 Q Which job will pay Danielle most money?

◉ Exam ready!

In this section you will:

- ◉ find out how fractions, decimals and percentages can be used to find solutions to practical problems
- ◉ apply a range of calculation techniques using fractions, decimals and percentages
- ◉ use checking techniques to confirm that solutions are right
- ◉ interpret and communicate solutions to practical problems, using calculations to justify any statements made
- ◉ round to an appropriate number of decimal places to solve practical problems

🔍 Take a look: Cake stall

Every year Karen makes cakes to sell at a charity fair.
Here is some information about Karen's cakes last year.

Number of cakes sold	Price charged per cake
50	£4

Karen thinks that more people will come to the fair this year.
She assumes she will sell 10% more cakes.
Karen wants to get at least 20% more money than she got for charity last year.
Karen needs to decide how much to charge for each cake at this year's fair.

Q **a)** Work out the minimum amount of money Karen wants to get from selling cakes this year.

b) Decide how much money Karen should charge for each cake.

Here's a possible solution:

a) $50 \times £4 = £200$ •————————• Ⓐ Work out how much money Karen got at last year's fair

$20\% \text{ of } £200 = 0.2 \times £200 = £40$ •————• Ⓐ Work out how much more money Karen wants to get this year

$£200 + £40 = £240$ •————• Ⓘ This is the minimum amount Karen wants to get this year

b) $10\% \text{ of } 50 = 5$ •————————• Ⓡ Work out how many cakes Karen thinks she will sell this year

$50 + 5 = 55$ •————• Ⓘ This is the number of cakes Karen thinks she will sell this year

$240 \div 55 = 4.363636\ldots$ •————• Ⓐ Divide the amount of money Karen wants to get by the number of cakes she thinks she will sell

Karen should charge £4.50 per cake. •————• Ⓘ Round up to a sensible price for a cake

Have a go

12 A newspaper article states that about 40% of school leavers in the UK go to university.

The head teacher of Middleton High School wants to know how her school leavers compare with the UK average of school leavers going to university.

Here are the school's records for school leavers last year.

Middleton High School – destination of school leavers

University	Employment
60	180

Ⓠ **a)** What fraction of the school leavers from Middleton High School went to University?

b) How does this information compare with the UK average for school leavers? Show all your working.

13 For his maths project, Ali asks 24 students to fill in a questionnaire.

For one question, $\frac{1}{2}$ the students answered 'yes',

$\frac{1}{4}$ of the students answered 'no'

$\frac{1}{4}$ of the students answered 'don't know'.

This is how Ali presented this information.

50%	25%	25%

Key:

■ Answered yes

□ Answered don't know

▨ Answered no

Ali wants to show the answers to each question in the same way.
Here are the answers for another question. Nobody answered 'yes'.

Answered 'don't know'	8
Answered 'no'	16

Q Draw a similar chart 9 cm long to show the answers to this question.
Include the percentage value for each answer.

14 Emma owns a restaurant.
Her customers can order **either** a snack **or** a lunch.
The table shows the prices of these meals.

	Average price per order
Lunch orders	£12
Snack orders	£5

During the week, about 10% of customers order lunch.
At the weekend, half the customers order lunch.

The table shows the number of customers one week.
Emma assumes that the number of
customers and the number of orders will
be about the same each week.

	Average number of customers each day
Monday to Friday	30
Saturday and Sunday	60

Q **a)** Work out the total number of customers Emma thinks will order lunch and snacks:
- from Monday to Friday
- on Saturday and Sunday.

b) How much money does Emma's restaurant get from lunch and snacks each week?

Now you can:

- Extract decimal, fraction and percentage values from questions in order to solve practical problems
- Add and subtract decimals
- Round amounts of money or measurements as appropriate
- Find common fractions or percentages of quantities to solve practical problems
- Order and compare common decimal, fraction and percentage equivalences to solve practical problems
- Use the answers you get to make sensible conclusions

3 Ratio and proportion
Know Zone

Ratios are useful in everyday problem solving. Although they can seem confusing, they are really quite easy. You need to understand basic facts about ratios and proportions.

Ratio

- A ratio is a way of comparing two quantities or more than two quantities.

  Example *George has £1 and Mike has £3. The amount of money George has compared with the amount of money Mike has can be written as the ratio 1:3*

- You usually simplify ratios as far as you can.
  To simplify a ratio, find a number that divides into both sides of the ratio.

  Example *15:60 simplifies to 1:4 (by dividing both sides by 15). Make sure you divide **both** sides by the largest number you can.*

- The order of the numbers in a ratio is important.

  Example *3:1 is not the same as 1:3*

- Ratios do not have units.

  Example *1:5 could mean 1 cm to 5 cm or £1 to £5; it depends on the context of the problem.*

- When comparing quantities that have units, make sure they are in the same units.

  Example *To simplify 5 km to 500 m, first convert 5 km to metres; the ratio becomes 5000:500 which simplifies to 10:1*

- The scale on a drawing or map is also a ratio.

  Example *A scale of 1:1000 may be shown on a map.*

- You can use a calculator to write a ratio in its simplest form.
  Find the fraction key on your calculator [$a\frac{b}{c}$ or ▤]

  Example *To simplify 12:36, key in 12 $a\frac{b}{c}$ 36 =
  The answer appears as* $\boxed{1\lrcorner3}$.

  Take care! If you key in ratios such as 200:50, the answer will appear as 4 only; remember to write it as a ratio, i.e. 4:1

Proportion

- Two quantities are in direct proportion if one quantity changes in the same way as the other quantity increases or decreases.

  Example If 1 taxi can carry 4 passengers, 3 similar taxis can carry 12 passengers.

- Two quantities are directly proportional if one of the quantities is a constant multiple of the other quantity.
  You must multiply or divide each quantity by the same number.

  Example The numbers of taxis and passengers in the above example have both been multiplied by 3.

Common errors

Some common errors when calculating ratios and proportion include:

- forgetting to convert to the same units, e.g. writing 2 m to 50 cm as 1:25

- making errors when converting units: make sure you know how many centimetres there are in a metre or millilitres in a litre

- ignoring the order of ratio in a question

- dividing or multiplying by the wrong number when using ratios or proportions

- increasing quantities using proportional relationships by adding instead of multiplying

⊙ Let's get started

In this section you will:
- ⊙ use and simplify ratios to solve practical problems
- ⊙ scale quantities that are directly proportional

🔍 Take a look: Making scones

In a recipe for scones, you need 50 g of butter to make four scones.

Q How many grams of butter do you need to make 36 scones?

> **R** The amount of butter you need is directly proportional to the amount of scones you make

💡 Here's a possible solution:
50 g   4 scones
? g   36 scones
36 ÷ 4 = 9

> **A** Write down 'what you have and what you want to find'

> **A** Divide 36 by 4 to see by how many times to multiply the number of grams of butter

Multiply 50 g by 9 = 450 g of butter.

🔍 Take a look: Investing in a business

Two friends set up a business together.
Jack invested £1500.
Brian invested £4500.
They share their profits in the same ratio as the money they invested.

Q Write this ratio in its simplest form.

💡 Here's a possible solution:
1500:4500

1500 ÷ 1500 = 1
4500 ÷ 1500 = 3

In its simplest form, the ratio is 1:3

> **R** The money they each invested is written in pounds so you can write £1500 and £4500 as a ratio straight away

> **A** You don't have to divide by 1500 straight away. Try dividing by 100 first; this gives 15:45.
> Then divide by 5 to give 3:9, and so on. You still end up with 1:3

Have a go

1   Jackie is training for a charity fun run.
    This is her training plan:

    walk   10 minutes
    run   2 minutes
    walk   5 minutes
    run   2 minutes
    walk   5 minutes

She wants to compare the time she spends walking with the time she spends running.

Q **Work out the ratio of Jackie's walking time to running time.**

2   A hairdresser has to mix hair dye from a base colour and copper colour in the ratio 1:2
    She mixes 27 ml of base colour with 9 ml of copper colour.

Q **Is she correct?**

3   Bronwen knows that if she walks for a mile she burns 100 calories.
    She wants to burn 1200 calories.

Q **How many miles does Bronwen need to walk?**

4   The instructions on the back of a bottle of weedkiller say:

    Mix 1 capful of weedkiller with 2 litres of water.

    Simon needs approximately a bucketful of weedkiller.
    The bucket holds a bit more than 9 litres.

Q **How much weedkiller does Simon need?**

5   Betty wants to find a part-time job.
    She sees these two job adverts.

Office cleaner
4 mornings a week
Wage: £24 per morning
(4 hours per morning)

Shop assistant
22 hours per week
Wage: £5.50 per hour

    Betty is happy to do either job.

Q **Compare the wages for each job.**
    Which job should Betty choose?
    Explain your reasons.

6   Ben and Nicky are walking in the hills.
   The weather is very bad so they want to find a shelter.
   There is a shelter marked on the map.
   The map has a scale of 1:2000
   The distance on the map is 5 cm.

Q How far must they walk to get to the shelter?
   Show all your working.

Fact

1 m = 100 cm.

⦿ We're on the way

In this section you will:
- ◎   use proportional relationships
- ◎   use ratios to find quantities
- ◎   justify your answers

🔍 Take a look: Shoe sales

The manager of a shoe shop keeps records of monthly shoe sales. The bar chart shows sales figures for sandals over the past three months.

The manager needs to decide whether to order some more sandals to sell in the next three months.

Sales figures for sandals

(Bar chart: *Number of pairs sold* on y-axis, 0 to 50; *Month* on x-axis. April = 40, May = 30, June = 10.)

Q **a)** Use the bar chart to find June's sales figures as a fraction of April's sales figures.

To help her decide, she compares the sales figures for April and June. If the sales figures for June are more than $\frac{1}{3}$ of the sales figures for April, the manager will order more sandals. If they are not, she will not order more sandals.

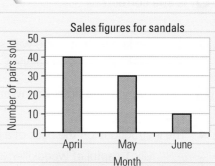

b) Does the manager need to order more sandals?

💡 Here's a possible solution:

a) In April 40 pairs of sandals were sold. •⎯⎯⎯⎯
   In June 10 pairs were sold.

   $\dfrac{10}{40} = \dfrac{1}{4}$ •⎯⎯⎯⎯

A Find the sales figures for June and April

A Express 10 as a fraction of 40

b) $\frac{1}{4}$ is less than $\frac{1}{3}$ so the manager does not need to order more sandals. •⎯⎯⎯

I Decide if $\frac{1}{4}$ is less than $\frac{1}{3}$

🔍 Take a look: Making fruit drink

Sandra is going to make a fruit drink for a children's party.
The drink is made by mixing 1 part squash with 5 parts water.
Sandra estimates that she needs 3 litres of fruit drink for the party.
She buys a half-litre bottle of squash.

Q **Has Sandra bought enough squash to make 3 litres of fruit drink?**
Show all your working.

💡 Here's a possible solution:
1 part + 5 parts = 6 parts
3 litres ÷ 6 = 0.5 litres.
0.5 litres is $\frac{1}{2}$ a litre.
Yes, Sandra has bought enough squash.

R '1 part squash with 5 parts water' means the ratio of squash to water is 1:5

A 6 parts needs to equal 3 litres. Squash is 1 part out of 6 parts. This means the amount of squash is 3 litres ÷ 6

🎯 Have a go

7 Jude and Jon are planning a Spanish holiday.
Here are the costs of staying at two Spanish hotels.

| Hotel Los Pelicanos | £540 per person for 10 nights |
| Hotel Las Ocas | £413 per person for 7 nights |

Q Which is the more expensive hotel?
Show all your working and explain your reasoning.

8 Doris sells hot drinks on a refreshment stall.
She must have enough tea for 500 people.
The tea is sold in cups which contain 300 ml of liquid when full.

Q **a)** How many litres of tea does Doris want to sell?

Doris assumes that every customer will take milk.
She also assumes that the ratio of milk to tea in each full cup is 1:5

b) Work out how many 2-litre cartons of milk Doris should buy.

9 Ali owns two clothes shops.

He wants to order jackets to sell in the second shop.

Here are last month's sales figures for jackets in the first shop.

Number of leather jackets sold	Number of denim jackets sold
16	96

Q **a)** Work out the ratio of leather jackets sold to denim jackets sold.

Ali decides that the number of leather and denim jackets he orders to sell in the second shop should be in the same ratio.

He orders 90 denim jackets to sell in the second shop.

Q **b)** How many leather jackets should Ali buy to sell in the second shop?

Exam ready!

In this section you will:

- use information in questions to write ratios
- identify and use ratios to find solutions
- find and use proportional relationships in order to scale quantities up or down
- calculate missing values in proportional relationships
- interpret answers to calculations
- state any assumptions you made in order to support your answers

Take a look: Visiting castles

The tourist board says 'there are on average three times as many visitors to Edinburgh Castle as there are to Conway Castle each year'.

Last year there were 2400 visitors at Edinburgh Castle.

Q **a)** Work out the likely number of visitors at Conway Castle last year.

The table shows entrance fees for each castle.

	Conway Castle	Edinburgh Castle
Entrance fee for one adult	£6	£11
Entrance fee for one child	£5	£5.50

About 3 out of every 5 visitors to both castles were adults.

b) Compare the money made from entrance fees at Conway and Edinburgh Castles last year.

A This problem is different from earlier problems: you are not given the total number of visitors, just the number at Edinburgh Castle

💡 Here's a possible solution:

a) $2400 \div 3 = 800$ visitors at Conway Castle. •————

> **R** The ratio is 3:1. 2400 represents 3 parts, not 4

b) Edinburgh: $\frac{3}{5} \times 2400 = 1440$ adults
$2400 - 1440 = 960$ children

Conway: $\frac{3}{5} \times 800 = 480$ adults
$800 - 480 = 320$ children •————

> **A** Work out the number of adults and children who visited each castle

Edinburgh: Adults $1440 \times £11 = £15\,840$
Children $960 \times £5.50 = £5280$
Total income $= £21\,120$

Conway: Adults $480 \times £6 = £2880$
Children $320 \times £5 = £1600$
Total income $= £4480$ •————

> **A** Work out the money from the entrance fees for each castle

$£21\,120 - £4480 = £16\,640$ •————

> **I** Compare the money made at each castle

The money made from entrance fees at Edinburgh Castle
was £16 640 more than the money made at Conway Castle.

🔍 **Take a look: Choosing a map scale**

Matt and Emma are going on a 5 km walk.
They want to buy a map of the area where they will walk.
They do not want the walk route **on the map** to be more than 50 cm.
They can buy maps with these scales: 1:5000 or 1:25 000

Q **a)** How many centimetres is 5 km represented by on each map?
b) Which map should Matt and Emma buy?

💡 Here's a possible solution:
a) $5 \text{ km} = 5 \times 1000 \text{ m} = 5000 \text{ m}$
$5000 \text{ m} = 500\,000 \text{ cm}$

> **I** First convert 5 km into centimetres. You can then apply each map scale

$500\,000 \div 5000 = 100 \text{ cm}$ •————

> **R** A map scale of 1:5000 means that every 1 cm on the map represents 5000 cm on the walk

$500\,000 \div 25\,000 = 20 \text{ cm}$ •————

b) 100 cm is more than 50 cm, 20 cm is less.

Matt and Emma should buy the map with
scale 1:25 000

> **I** On the 1:5000 scale map, 5 km is represented by 100 cm. On the 1:25 000 scale map, 5 km is represented by 20 cm

> **I** Decide which map fits Matt and Emma's needs

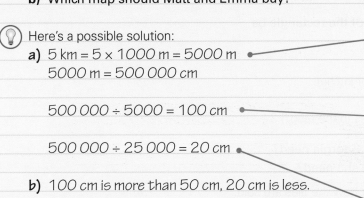

Have a go

10 Jill is planning refreshments for a charity fair. There will be about 800 people at the fair. Jill assumes that everyone will buy a cup of tea or a cup of coffee.

Jill thinks that three times more people will buy tea rather than coffee.

(Q) a) How many cups of tea will Jill need to make?

She knows that she can make two cups of tea from one tea bag. Tea bags are sold in boxes of 160.

b) How many boxes of tea bags should Jill buy?

11 Rosita pays her gas bill every three months.
For winter her gas bill was £189.
Rosita estimates these ratios for her gas usage:
  Winter usage to Spring usage = 2:1
  Winter usage to Autumn usage = 2:1
  Winter usage to Summer usage = 3:1

(Q) Use this information to estimate:

a) Rosita's gas bills for the summer.

b) The total amount Rosita is likely to pay for gas in a one-year period.

12 Alan is designing a new kitchen. He needs to draw a plan of the kitchen floor.
He wants to draw the plan to a scale of 1:20
Alan's kitchen floor is a rectangle, 4 m by 6 m.

Think First!

1 m = 100 cm

(Q) What will the measurements of the rectangle be on the scale plan?

13 Georgia plans to tile a wall in her kitchen.
She needs a total of 1536 tiles.
She will use black and white tiles in the ratio 1:3

(Q) a) How many black tiles does Georgia need?
How many white tiles does she need?

Here is some information about how the tiles are sold.

White tiles	Packs of 20
Black tiles	Packs of 10

(Q) b) How many packs of each colour of tile should Georgia buy?

14 Brian is buying carnations to plant in his garden.

Brian wants to plant only red and white carnations.

Brian wants to plant at least 5 red carnations for every white carnation.

He estimates that he can plant about 60 carnations in his garden.

Here is some information about the carnations.

Boxes of 6 red carnations £6	Boxes of 6 white carnations £5

Q How much will it cost Brian to buy the carnations he wants?

Show all your working and explain all your reasoning.

Now you can:

- Use information to find ratios and proportions
- Use ratios to divide quantities and use proportions to scale quantities up or down
- Evaluate what the answers you get from using ratios and proportions mean in the context of the question

In the exam there will be questions that need you to work with different features. You will need to process the information, and make comparisons, in order to find an effective solution.

ResultsPlus
Maximise your marks

Question

Josh makes bookings for people to rent rooms for meetings and events.
This table shows the cost to rent rooms of different sizes.

	Recommended number of people	Per hour	Per day
Room A	100–119	£14	£100
Room B	120–180	£16	£130
Room C	160–180	£18	£160

This table shows some information about the number of people needing rooms from Monday to Friday.

	Number of people	Time room is needed (hr)
Monday	100	6
Tuesday	120	8
Wednesday	170	10
Thursday	110	12
Friday	150	10

Josh wants to charge each group of people the smallest possible rent.

Work out the cheapest rent of a room for each group of people.

Examiner tips

First, find out the rooms needed on each day. Then compare the costs of renting the rooms.

Student response	Examiner comments
Let's look at a poor answer: It is cheaper for Josh to book Room B for £128 than for £130.	This answer compares the hourly and daily rate for Room B. However, the answer does not consider which room to choose for each day. No calculations are shown.
Here is an answer that would gain a majority of the marks: 100 people room A Per day £100, per hour £14 × 6 = £84 120 people: Room B Per day = £130, per hour £16 × 8 = £128 170 people: Room C Per day = £160, per hour £18 × 10 = £180 110 people: Room A Per day £100, per hour £14 × 12 = £168	This answer has some good features: • It considers the best choice of room for each day. • It compares the hourly and daily rates. However, there are some errors and some things are missing. *Error* *On Wednesday Room B is cheaper to book than Room C.* *Missing* *There is no processing for Friday.* *The cheapest options need to be described.*

Student response	Examiner comments
Let's see what we could do to get full marks: Room A: Monday, Thursday Room B: Tuesday, Wednesday, Friday Mon: 14 × 6 = £84 cheaper to rent by the hour Tue: 16 × 8 = £128 cheaper to rent by the hour Wed: 16 × 10 = £160 cheaper to rent by the day Thu: 14 × 12 = £168 cheaper to rent by the day Fri: 10 × 16 = £160 cheaper to rent by the day Total = £572	This answer addresses all of the conditions of the question. It will gain full marks. • Correct room types are chosen for all days. • It calculates and compares in order to find the cheapest choice of rooms.

ResultsPlus
Exam Question Report

Exam question

A customer sent this order to a farm shop.

> **500 g of beef at £7.80 per kg**
>
> **1 packet of charlotte potatoes at £1.00**
>
> **1 packet of mini carrots at 90p**

Calculate the total cost of this order.　　　　　　　　　**(4 marks, Jan 2010)**

How students answered

66% of students (0–1 mark)	2% of students (2 marks)	32% of students (3–4 marks)
Most students could not work out the cost of the beef. The key points are remembering that 1000 g = 1 kg and £7.80 ÷ 2 = £3.90	Hardly any students made errors in showing information.	Some students did not correctly add together the amounts of money. They made a mistake converting between units.

Put it into practice

1. Find some data about renting things from a hire company, such as for DIY equipment, cars or vans. Or, look at hotel room rates given in days and weeks.
   Discuss with a partner how the renting costs are calculated.

2. Find food recipes on the internet or in a cookery book. Choose a recipe. Calculate the actual cost of buying the ingredients.

ASSESSMENT PRACTICE 1

1 Jenna has made 200 toys. She is going to put the toys into boxes.
She will put 10 toys in each box.
Jenna says she will need 2000 boxes.

Jenna is wrong.

(Q) a) Explain why.

Jenna sells the 200 toys to a shop. The shop pays Jenna £75 for each box of toys.

b) What is the total amount of money the shop pays Jenna?

2 Gavin is going to make egg salad for his daughter's birthday party.
Here is part of the recipe for 6 people:

> 2 tablespoons of grated cheese
> 3 eggs
> 45 ml salad dressing

Gavin is going to make the egg salad for 30 people.

(Q) a) How many eggs will he need?

Gavin needs a bottle of salad dressing to make enough egg salad for 30 people.
Salad dressing is sold in different sized bottles.

Salad dressing bottle	Size
Small	150 ml
Medium	250 ml
Large	300 ml

b) What size bottle of salad dressing does Gavin need?
Explain why.

3 A company makes green paint by mixing three parts yellow paint to one part blue paint.

On one day, the company made 2500 litres of green paint.

(Q) How much yellow paint did they need?

4   Shelf is an oil drilling company. Shelf has calculated the cost of starting to drill at four sites.

Site	Cost of preparation (£ millions)
A	6.7
B	3.2
C	5.5
D	1.9

Shelf wants to drill at as many sites as possible.

It has a budget of £13.2 million.

Q Which of the sites can Shelf start to drill at?

5   Kevin works in a supermarket. He is checking an advertising leaflet to make sure that the information is correct.

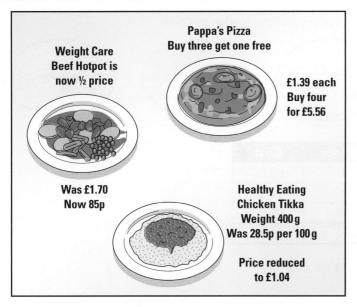

Q What corrections does Kevin need to make?

6   The cost of a second-hand car depends on its mileage.

Boyce buys old second-hand cars. He uses this table to work out the value of a car.

Mileage	Value
Low	25% of cost when new
Medium	10% of cost when new
High	5% of cost when new

Derek wants to sell his car.

The car has medium mileage. The car cost £20 000 when new. Boyce is going to buy the car.

Q How much money will Boyce pay Derek?

7 The table shows the lengths of four rivers to the nearest kilometre.

River	Length to the nearest km
Dee	113
Thames	346
Severn	354
Trent	290

Q a) Write down the lengths of the rivers to the nearest 100 km.

Kim wants to canoe the total length of a river.
She wants to canoe a river that is longer than 150 km, but **not** longer than 350 km.

b) Which rivers can Kim choose?

8 Xi goes to a sushi bar.
The colour of each dish shows the price of the food in the dish.

Dish colour	Cost
White	£3.50
Grey	£4.00
Green	£4.00
Yellow	£4.50

Xi wants to buy two different dishes of food.
He will spend exactly £8.

Q Write down all the different pairs of dishes that Xi can choose.

9 The table shows the cost for an adult staying in a hotel for one week.

Holiday start date	Cost
1 June–1 July	£400
2 July–15 July	£550
16 July–3 September	£800
4 September–23 October	£400

Child
50% *off*
adult cost

Marie and John and their three children want to stay in the hotel for one week in the school holidays (12 July–3 September).

They have a budget of £2000.

Q On what dates can this family have a week staying in the hotel?

10 Cynthia works for an agency. She cleans hotels in central London.

The agency pays Cynthia a basic rate of £6.50 per hour.
The agency pays her £7.50 an hour for each hour of overtime
she works.

Cynthia kept a record of the hours she worked last week.

Day	Hours Basic	Hours Overtime
1	6	
2	$8\frac{1}{2}$	
3		$3\frac{1}{2}$
4	4	
5	9	
6		$3\frac{1}{2}$
7		

The agency paid Cynthia £203.75 for the work she did last week.

The agency did not pay Cynthia the correct amount of money.

Q **How much money does the agency owe Cynthia?**

4 Time
Know Zone

You do time calculations every day. Think about how often you check your watch or look at a timetable. Even working out how long it is until your favourite TV programme is a time calculation!

Units of time

First you need to know some key facts about time:

60 seconds	=	1 minute	52 weeks	=	1 year
60 minutes	=	1 hour	12 months	=	1 year
24 hours	=	1 day	10 years	=	1 decade
7 days	=	1 week	100 years	=	1 century

A time can be written in different ways. You can write a time using the 12 hour clock or using the 24 hour clock.

12 hour clock		24 hour clock	
OK	Wrong	OK	Wrong
08:30 am	830	17:20	720
8.30 am	8.3 am	17.20	7.2
		1720	1720 pm

- Twenty past nine can mean 0920 or 2120
- Twenty-five to seven can mean 0635 or 1835
- Quarter to three can mean 0245 or 1445

A time **interval** is the amount of time from one time to another.
You can also write a time interval in different ways.

OK	Wrong
3 hours 20 minutes	3.2
3 hr 20 min	3.2 hours
3 20	3 h 2
200 minutes	3.2 min

Common intervals of time are used in everyday speech:
- Quarter of an hour = 15 minutes
- Half an hour = 30 minutes
- Three-quarters of an hour = 45 minutes

Let's get started

In this section you will:
- add together lengths of time
- write time unambiguously

🔍 Take a look: Deadlines

Jenny told a customer that the garage would repair his car on Tuesday of the next week.

Jenny said this on Monday 25 January 2010.

Q What was the date of Tuesday of the next week?

💡 Here's a possible solution:

Tue 26 Jan, Wed 27 Jan, Thu 28 Jan, Fri 29 Jan,
Sat 30 Jan, Sun 31 Jan, Mon 1 Feb, Tue 2 Feb.
The date of Tuesday of the next week was 2 February.

R January has 31 days – list the days until Tuesday of the next week

🔍 Take a look: Bus times

Osman catches a college bus at twenty to four.
He gets off the bus $\frac{3}{4}$ of an hour later.

Q What time does Osman get off the bus?

💡 Here's a possible solution:

Osman catches the bus at 3.40 pm.
$\frac{3}{4}$ of an hour is 45 minutes.

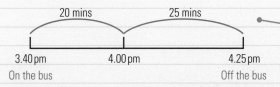

20 mins   25 mins

3.40 pm   4.00 pm   4.25 pm
On the bus   Off the bus

A We need to add 45 minutes to 3.40 pm

Osman gets off the bus at 4.25 pm.

🎯 Have a go

1 The table below shows part of a train timetable.

Kings Cross	1300
York	1534
Durham	1625
Edinburgh	1825

(Q) What is the journey time from Kings Cross to Durham?

2 A train leaves Edinburgh at five minutes to three.
The train arrives in Glasgow at five minutes past four.

(Q) What is the journey time from Edinburgh to Glasgow?

3 Darren works in a factory.
The factory will close for two weeks on 27 June.
Darren wants to go on a 10-day holiday when the factory is closed.

(Q) Between which dates can Darren go on holiday?

4 On Wednesday 27 June Emily receives a letter.
The letter asks Emily to go for a job interview on 5 July.

(Q) On which day of the week is Emily's job interview?

5 Mrs Weaver is having her house decorated.
The decorators tell her that the job will take 12 days.
The decorators do not work at the weekend.
The decorators start work on Monday 24 March.

(Q) What day and date will they finish the job?

6 Mario watches a film.
The film is 2 hours and 25 minutes long.
The film starts at 1840

(Q) What time will the film finish?

7 Bella and Mike plan a day out.
They will leave home at 0930 to drive to Brighton.
They expect that the journey will take 1 hour and 45 minutes.

(Q) What time do they expect to arrive in Brighton?

We're on the way

In this section you will:
- ◉ subtract periods of time
- ◉ communicate decisions

Take a look: Cooking a meal

Dave is going to cook a meal. He needs to serve the meal at 8 pm. The meal will take 2 hr 40 min to cook. The meal also needs 15 min to cool down when Dave takes it out of the oven.

Q What is the latest time that Dave should put the meal in the oven?

Q Here's a possible solution:

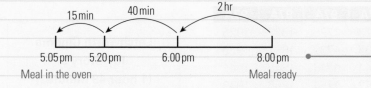

15 min	40 min		2 hr
5.05 pm	5.20 pm	6.00 pm	8.00 pm
Meal in the oven			Meal ready

A We need to subtract time from 8 pm
8 pm − 2 hr = 6 pm
6 pm − 40 min = 5.20 pm
5.20 pm − 15 min = 5.05 pm

Dave should put the meal in the oven by 5.05 pm at the latest.

Have a go

8 A plane leaves Gatwick airport in the UK. The plane arrives at Johannesburg airport in South Africa at 8 am.

- The flight time is $11\frac{1}{2}$ hours.
- The time difference between the UK and South Africa is +2 hours.

Q What time did the plane leave Gatwick airport?

9 Bob has to finish building a garage on or before Thursday 7 June 2012.
Bob thinks it will take him 14 working days to build the garage.
He works from Monday to Thursday inclusive.

Think First!
You need to use your knowledge of the number of days in May to help you solve this problem.

Q What is the latest date Bob could start building the garage?

Exam ready!

In this section you will:
- ⊙ coordinate features in solving a situation
- ⊙ explain your answer clearly

Take a look: Going to the cinema

Laura lives in Stotfold. She is planning to go to the cinema in Letchworth.

Her mother can take Laura to the cinema any time after 2 pm.
Laura will have to get the bus home.

Laura will catch the bus home from Letchworth to The Green, Stotfold.
The last bus leaves Letchworth at 1925
It takes 5 minutes to walk to the bus station from the cinema.

Bus timetable

	97A	97A	97A	97A
Letchworth	1623	1725	1827	1925
Grange Estate	1637	1739	1841	1925
Fairfield Park	1644	1746	1848	1941
Stotfold, High Street	1652	1754	1856	1949
Stotfold, The Green	1654	1756	1858	1951

> **Letchworth Cinema**
> **Toy Story 3**
> **(1 hour 49 minutes)**
> Showing at: 1520   1745

Q Plan a schedule for Laura's visit to the cinema.

💡 Here's a possible solution:

Film starts	15 20
Film ends	17 09
Bus from Letchworth	17 25
Arrive in Stotfold	17 56

A Laura cannot go to the film that starts at 1745 as this will finish at 1934, which is after the last bus has left.

R Write out the schedule clearly showing all the necessary information.

Have a go

10 A nursery needs one adult for every eight children.
The nursery has seven members of staff: Mr Allen, Miss Baker, Mr Case, Miss Dole, Mrs Egan, Mrs Foden and Miss Green.

Each member of staff works for five days.
The nursery is open for six days.

Q **a)** Work out a rota for the staff.
b) How many children can attend the nursery each day?

11 There are three car parks near to a city centre.

Car Park A	Open 6 am–8 pm – 7 days a week	
	Up to 1 Hour	£1.00
	Up to 2 Hours	£2.00
	Up to 3 Hours	£3.00
	Up to 4 Hours	£4.00
Car Park B	Open 24 Hours Monday – Sunday including Bank Holidays	
	0–2 Hours	£3.00
	2–4 Hours	£4.60
	4–6 Hours	£7.20
Car Park C	Open 24 hours a day including Sundays	
	Up to 1 Hour	£2.00
	Up to 2 Hours	£3.40
	Up to 4 Hours	£4.80

Geoff wants to park near to the city centre.
He wants to park at 6 pm.
He will get back to his car $3\frac{1}{2}$ hours later.

Q a) Which car parks are open when Geoff wants to park?

b) Which car park will be the cheapest for Geoff to use?

12 A cinema has the following programme for Wednesday. The programme shows the start times of the films and the length of each film.

Adeleke and her partner are planning to go to the cinema. They want to see two different films on Wednesday. They want to leave the cinema by 6.30 pm.

Q Write down all the possible options for Adeleke and her partner to watch **two** films.

3D Avatar (12A)	**Invictus** (15)
162 min	133 mins
12:00 15:50 19:50	20:10
A Single Man (12A)	**Leap Year** (PG)
101 min	100 mins
14:20 17:20 20:20	13:00 15:30
	18:00 20:30
From Paris with Love (15)	**Valentine's Day** (PG)
92 min	124 mins
13:30 16:00	14:10 17:00 20:00
18:30 21:00	

Now you can:

- Add time and subtract times
- Use time periods in situations
- Use time in various contexts and situations to solve problems

5 Measures
Know Zone

You use many different forms of measure in everyday life, from temperatures in weather forecasts to your own weight and height. You need to know which units are used for temperature, length, weight and capacity and how to calculate with them.

Temperature

Temperature is usually recorded in degrees Celsius (°C).
Sometimes you may see degrees Fahrenheit (°F) used.
0°C is the approximate melting point of ice. 100°C is the approximate boiling point of water.

Length

Length is a measure of distance. The most common units used to measure length are:

Metric		Imperial	
millimetres (mm)	centimetres (cm)	inches (in)	feet (ft)
metres (m)	kilometres (km)	yards (yd)	miles (mile)

Weight

Weight is a measure of how heavy an object is. The most common units used to measure weight are:

Metric			
milligrams (mg)	grams (gm)	kilograms (kg)	tonnes (t)
Imperial			
ounces (oz)	pounds (lb)	stones (st)	tons (t)

Capacity

Capacity is a measure of volume. The volume of a container is the amount of space inside it. The capacity of a container is the amount of fluid it can hold. The most common units used to measure capacity are:

Metric		
millilitres (ml)	centilitres (cl)	litres (l)
Imperial		
fluid ounces (fl oz)	pints (pt)	gallons (gal)

Metric units

The metric system is based on tens, hundreds and thousands. You need to know how to convert between the metric units for length, weight and capacity.

Length	Weight	Capacity
10 mm = 1 cm	10 mg = 1 g	10 ml = 1 cl
100 cm = 1 m	1000 g = 1 kg	100 cl = 1 litre
1000 m = 1 km	1000 kg = 1 tonne	1000 ml = 1 litre

Imperial units

Imperial units are still used in the UK. For example, distances on road signs are given in miles. Drinks can be measured in pints. You should be able to work with the connections between imperial units.

Mileage charts

You can use a mileage chart to find the distance between two places.

Aberdeen	Cardiff	Manchester	London
529			
345	201		
549	150	201	

The distance between Cardiff and London is 150 miles.

Estimating measure

It is useful to estimate measure. Here are some examples.

Example A standard bag of sugar weighs 1 kg (approximately 2 lb).

Example A standard coffee mug holds about 250 ml or $\frac{1}{4}$ litre.

Example The average height of a male in the UK is about 1.8 m or 5 ft 10 in.

Example The width of your little finger is about 1 cm.

Let's get started

In this section you will:
- read and use scales
- use scales to estimate, measure and compare length, distance, weight, capacity and temperature

🔍 Take a look: Using scales

A nurse measures a patient's temperature in degrees Celsius.

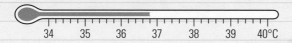

Q What temperature does the thermometer show?

A First work out what each mark on the scale represents

Here's a possible solution:

5 divisions on the scale represent 1°C, so 1 mark represents $\frac{1}{5}$ = 0.2°C.

The temperature is one mark below 37°C so the temperature is 37 − 0.2 = 36.8°C.

🎯 Have a go

1   Eugene is making some chicken soup.
    First, she measures the amount of water she needs.

Q How much water is in the jug?

2   A scale marked in grams is used to weigh a parcel.

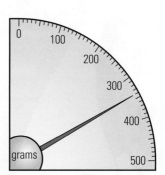

Q What weight does the scale show?

3   Stephanie marks her daughter's height on a chart on her daughter's fourth
    birthday and her fifth birthday.

Q Use the chart to work out how much Stephanie's daughter has grown
between her fourth and fifth birthdays.

4   Donna is feeling ill. The thermometer shows her temperature.

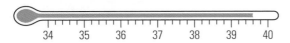

> **Fact**
>
> The human body's normal
> temperature is 36.8°C.

Q Donna's temperature is above normal body temperature.
How many degrees above?

5   Karen has been on a diet. She has lost 3.5 kg.
    The scales show Karen's weight now.

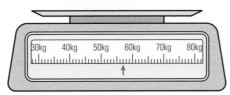

Q How much did Karen weigh when she started her diet?

6   The full fuel tank of Loris's car holds 40 litres of petrol.
    The fuel gauge shows how much petrol is in the fuel tank.

Q Estimate how much petrol is in the fuel tank.

7   The diagram shows the voltage across a circuit on a voltmeter.

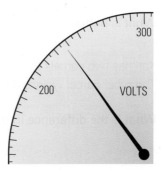

Q What is the voltage, to the nearest 10 volts?

We're on the way

In this section you will:
- convert units of measure within the same system
- use a mileage chart

Take a look: Working with capacity

Jeff buys orange juice in cartons of three of different sizes: 75 ml, 750 ml and 1.75 litres.

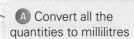

Q Can Jeff put all the juice into a 2.5 litre jug?

A Convert all the quantities to millilitres

Here's a possible solution:

First convert 1.75 litres to millilitres:
1.75 × 1000 = 1750 ml.

The total amount of juice is 75 + 750 + 1750 = 2575 ml.
The 2.5 litre jug holds 2500 ml.

So Jeff cannot put all the juice into a 2.5 litre jug.

Have a go

8 A tiling pattern is made from three small tiles.
The lengths of the tiles are 5.3 cm, 32 mm and 19 mm.

Q What is the total length of the three tiles in centimetres?

9 Bethany's weight is 8 stone 4 pounds.

Q What is Bethany's weight in pounds?

> **Fact**
>
> 1 stone = 14 lb

10 Karl has two parcels to post. One parcel weighs 4.5 kg.
The other parcel weighs 4.05 kg.

Q What is the difference in the weight of the two parcels?

> **ResultsPlus**
> **Exam Tip**
>
> It is a common mistake to write 4.5 kg as 4 kg 50 g. This is wrong because 1 kg = 1000 g.

11 Katie buys 3 kg of mixed sweets. She is going to sell bags of the sweets to raise money for charity. She puts 10 sweets into each bag. Each sweet weighs between 6 g and 8 g.

Q Estimate the number of bags of 10 sweets that Katie can make from the 3 kg of sweets.

12 One lap of an athletics track is 400 m.
Some athletes run a 10 km race.

Q How many times do the athletes run round the track?

13 Mrs Khan uses plastic drinking glasses for her daughter's party.
Each glass holds between 150 and 200 millilitres.
Mrs Khan has a 1.5 litre carton of apple juice.

Q Estimate the number of plastic glasses she can fill.

Take a look: Using mileage charts

Mileage charts show the distances in miles between cities or towns.
The figures in this mileage chart show the distances between four cities, in miles.

Birmingham	Edinburgh	Exeter	Liverpool
298			
164	454		
102	224	258	

Q Use the mileage chart to work out the distance between Birmingham and Exeter.

Here's the solution:

Birmingham	Edinburgh	Exeter	Liverpool
298			
164	454		
102	224	258	

R First find the column for Birmingham. Read down this column and find where it crosses the row for Exeter

R The value in this cell shows the distance between Birmingham and Exeter

The distance is 164 miles.

Have a go

The mileage chart below shows distances between towns and cities in the south west of England.
Use the chart to answer questions 14 to 18.

Barnstaple	Bristol	Exeter	Penzance	Plymouth	Taunton
100					
55	84				
108	194	110			
67	125	44	77		
50	51	34	144	75	

Q 14 What is the distance between Penzance and Exeter?

Q 15 Which two towns are 50 miles apart?

Q 16 a) What is the distance from Bristol to Taunton?

b) Bill drives from Bristol to Taunton. He then drives from Taunton to Penzance. How far does Bill drive altogether?

Q 17 Which two towns are closest together?

Q 18 Which two towns are furthest apart?

19 The distance in miles between:
- Edinburgh and Exeter is 454
- Edinburgh and Liverpool is 224
- Edinburgh and Manchester is 219
- Exeter and Liverpool 258
- Exeter and Manchester is 246
- Liverpool and Manchester is 35

Q Draw a mileage chart to show the distances between Edinburgh, Exeter, Liverpool and Manchester.

Exam ready!

In this section you will:
- ⊚ estimate measures
- ⊚ use measures to find solutions to problems
- ⊚ use measures to make judgements

🔍 Take a look: Estimating length

The photograph below shows a family looking at a statue of Peter Pan.

Fact

The average height of a man in the UK is 1.8 m

Q Estimate the height of the Peter Pan statue.

A Use the average height of a man. Work out how many times higher the statue is than the man and use this to calculate the height of the statue

💡 Here's a possible solution:

The statue is about 2 times the height of the man.
2 × 1.8 = 3.6 m

◉ Have a go

20 The picture shows a man standing next to a building.

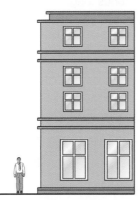

Q Estimate the height of the building.

21 James wants to estimate the length of a field.
He knows that each of his steps measures about 75 cm.
James takes 160 steps to walk across the field.

(Q) Work out an estimate for the length of the field.
Give your answer in metres.

22 Karl works for a communications company in Basingstoke.
He needs to travel to offices in Woking,
Reading and Swindon.

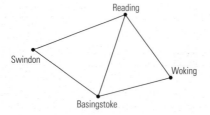

> **Think First!**
>
> Use the diagram to help you decide on a sensible possible route.

The table shows the distances between each of these towns in miles.

Basingstoke			
23	Woking		
13	20	Reading	
35	54	34	Swindon

Karl leaves his office in Basingstoke. He visits all the towns.
He then goes back to his office.

(Q) Plan Karl's shortest route.

23 Mia wants to lay some paving stones in her garden to make a path.

The path will be one paving stone wide and 12 paving stones long.
The paving stones she buys measure 43 cm by 43 cm.

(Q) **a)** Work out the length of Mia's path.

The tiles are sold in packs of 5.
Each pack costs £32.99

(Q) **b)** How much does it cost Mia to buy the paving stones?

24 Tasha buys a 5 litre container of ice cream and 20 ice cream cones.

Tasha's ice cream scoop has a volume of 134 cm^3.

Tasha wants to put two full scoops of ice cream in every ice cream cone.

(Q) Does Tasha have enough ice cream to put two full scoops in each ice cream cone?

Now you can:

- Read and use scales
- Use scales to estimate, measure and compare length, distance, weight, capacity and temperature
- Convert units of measure within the same system
- Use a mileage chart
- Use a conversion graph
- Estimate measure

6 Drawing and Measuring
Know Zone

Measuring accurately is an important skill. If you are planning where to put furniture in a room, it is easy to draw a scale diagram, but it's difficult to move heavy furniture about. It can even stop us making the expensive mistake of buying something that will not fit into the room!

Measuring

You need to know how to choose the right instrument and the correct scale.

This scale is in centimetres but you can also measure in millimetres

This scale is in millimetres

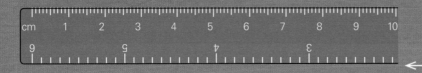

This scale is in inches

If you want to measure a curved shape, e.g. your waist, then a tape measure is a better choice.

Let's get started

In this section you will:

- choose the best instrument to measure with
- choose the best units to measure in
- measure accurately
- write your answers unambiguously

🔍 Take a look: Measuring length

Here is a picture of a dog.

Q Write down the height of this picture.

💡 Here's a possible solution:

R Find an approximate answer. (This will help you avoid making a silly mistake.) In this case the picture is about 5 cm high

R Make sure that the beginning of the scale is positioned exactly at the edge of the picture

R Read off as accurately as you can. On this ruler, the small divisions each represent 1 mm

The picture has a height of $5\frac{1}{2}$ cm = 5.5 cm.

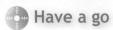 Have a go

1   Look at the picture of the dog on page 3.

(Q) Measure the width of the picture in centimetres.

2   Look at the picture of the pencil.

(Q) Measure the length of the pencil in
     centimetres.

(Q) 3 Measure the length and width of this text
       book in centimetres.

(Q) 4 Draw accurately a rectangle that measures 8 cm long and
       4.5 cm wide.

We're on the way

In this section you will:

- measure angles
- interpret information from a scale drawing
- choose the correct measuring equipment
- communicate results using the correct units

Take a look: Measuring angles

To measure angles we need an angle measurer
or a protractor. Angles are measured in degrees.

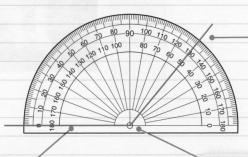

A Measure round from
0 clockwise (outer scale), or
from 0 anticlockwise (inner
scale)

A Put your protractor
along the base line

A Place the centre of the
protractor on the vertex of the
angle

The angle is 130°.

Alan has a scale drawing of part of a model boat. Alan wants to make the drawing larger.
He decides to measure the two smaller angles of the triangle.

Q Measure the two smaller angles of the triangle.

A Estimate the size of the angles first. Both the smallest angles are less than 90°. This should help you use your protractor accurately

Here's a possible solution:

A To measure the top angle line up the base of the protractor here at 0°

R Read off where this line crosses the scale. Make sure you are using the correct scale.
Your estimate should tell you it isn't 120°

I Remember to write the degrees symbol

The angles are 60° and 45°.

🔍 Take a look: Scale drawing

The diagram shows a ladder leaning against a wall. The diagram is drawn to scale.
Health and safety regulations state that the ladder should be at an angle of 75° to the ground.

Q a) How long is the ladder?
b) Does the position of the ladder meet the Health and Safety regulations?

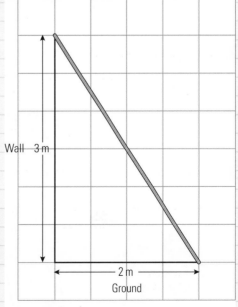

Wall 3 m

2 m
Ground

Scale: 1 cm = $\frac{1}{2}$ m

Here's a possible solution:

a) The ladder is 7.2 cm long.

So the real ladder is 7.2 × 0.5 m long = 36 m.

> **A** Measure the ladder using your ruler. Then use the scale to find the length of the real ladder

b) The angle with the ground is 56°.

So the position of this ladder does **not** meet the health and safety regulations.

Have a go

5   The diagram shows the ends of the roofs of three buildings.

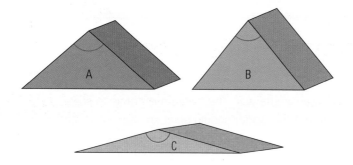

Q Measure the angle marked on each of the gable ends.

6   An articulated lorry can only turn around corners that have an angle greater than 70°.

Q Which of these corners can the lorry get round?

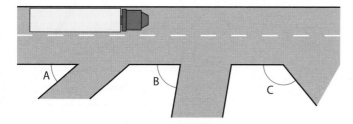

7 The diagram shows a ramp which Doria
wants to use to provide wheelchair access
to her Dress Shop.
The width of the pavement outside Doria's
Dress Shop is 4 m.

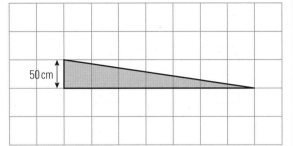

50 cm

Q **a)** Can this ramp be placed on the
pavement outside Doria's Dress Shop?

The maximum angle for a wheelchair ramp is 10°.

Q **b)** Can a wheelchair be pushed up this ramp?

8 A crane can lift a weight of 5 tons if the jib is at an angle of
135° or less than 135° to the horizontal.

Q Which of these cranes can lift the weight?

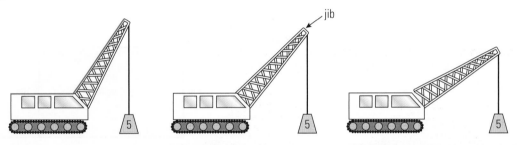

jib

◉ Exam ready!

In this section you will:

- ◉ represent 3D objects as a 2D plan
- ◉ use a scale drawing to position items
- ◉ communicate your results

🔍 Take a look: The office

Sue is designing an open plan office.
The top of each desk is 2 m by 1 m.
Sue has already decided where she will put the printer.
There must be at least a 1 m gap between each desk.
There must be at least a 1 m gap between a desk and
the door, and between a desk and the printer.

Q Draw a scaled plan of where the desks will be.
How many desks can Sue have in the office?

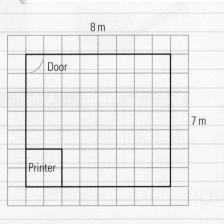

8 m

Door

7 m

Printer

Here's a possible solution:

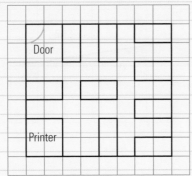

Sue can put nine desks in the office.

> **R** Recognise that two squares will represent one desk

> **R** Start 1 m away from the open door

> **I** Draw a clear diagram using a ruler. **Don't** forget to write down the result

Have a go

9 The diagram shows the dimensions of a garden.

Omel wants to dig a vegetable plot in the shape of a rectangle.
She wants the vegetable plot to be 3 m long and 2 m wide.

The vegetable plot must be at least 5 m away from the house.

10 m

6 m   Garden   House

Q Using a scale of 1 cm to 1 m, draw a diagram of the garden.
Show a possible position for Omel's vegetable plot.

10 Jill is designing her bedroom. The room is 2 m by 4 m.
Her room already has a wardrobe. She wants to add a desk and a bed.
She has a choice of three beds.

Desk: Length 120 cm, Width 80 cm

Bed	Width	Length
small	80 cm	165 cm
medium	90 cm	180 cm
large	135 cm	190 cm

Window

Fitted wardrobe

2 m

Door

4 m

> **Think First!**
> You will need to place the bed before you can add the desk.

Q Use Resource Sheet 6.1 to design Jill's bedroom.

11   The diagram shows the position of a small boat which has lost its power.
A lifeboat is sent to help. The diagram also shows the lifeboat.

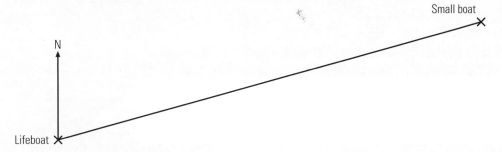

The scale of the diagram is 1 cm = 1 km.

(Q) **a)** How far is the lifeboat from the small boat?

The lifeboat is facing north.

(Q) **b)** Through how many degrees must the lifeboat turn to reach the
boat?

12   Mrs Smith wants to put solar panels on her roof.
The picture shows the south-facing roof of her house.
Mrs Smith wants to place as many solar panels as
possible on this side of the roof.

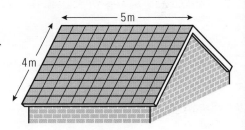

Each solar panel is a rectangle.
They measure 2 m by 1.5 m.

(Q) Using a suitable scale, draw a diagram of the roof
showing the solar panels. Use centimetre squared paper.

Now you can:

- Measure lines to the nearest millimetre
- Measure angles to the nearest degree
- Recognise line symmetry
- Represent real situations using scale drawings
- Interpret your scale drawings to solve real problems
- Use your skills in practical situations

Formulae are used to describe the relationship between two or more **variables**.
You can use this formula to calculate the fuel efficiency of a car:

$$\text{fuel efficiency (miles/gallon)} = \frac{\text{distance driven (miles)}}{\text{amount of petrol used (gallons)}}$$

Example So if a car uses 2 gallons of petrol to drive 100 miles, you can substitute these values in:

$$\text{fuel efficiency} = \frac{100}{2}$$

and do the calculation to get the solution:

fuel efficiency = 50 miles/gallon

Order of operations

You need to understand **BIDMAS**; you must always follow the correct order of operations.

B – Brackets
I – Indices
D – Division
M – Multiplication
A – Addition
S – Subtraction

Example The formula for final velocity is:
final velocity (m/s) = initial velocity (m/s)
+ acceleration (m/s^2) × time (s)
Calculate the final velocity when: initial velocity = 9 m/s;
acceleration = 5 m/s^2 and time = 20 s.
Final velocity = 9 + (5 × 20)
  = 9 + 100
  = 109 m/s

> You must do the multiplication first, then the addition

When you substitute into formulae, make sure that each variable is in the correct units.

Example The formula for speed is:

speed (km/hour) = distance (km) ÷ time (hours)

A car travels 30 000 metres in 2 hours.

Calculate the speed of the car.

Speed = 30 ÷ 2

= 15 km/hour

Always check that your answer is sensible. This can help you spot when you have used the incorrect units.

Let's get started

In this section you will:
- use the correct order of operations, BIDMAS
- think about the appropriateness of results
- set out solutions methodically

Take a look: Mechanics' costs

A garage uses the formula below to calculate how much to charge each customer:

amount charged (£) = cost of parts (£) + 20 × hours of work

Lauren takes her car to the garage to be fixed.
The parts that need replacing cost £79.
Two mechanics take three hours each to fix her car.

Q How much should the garage charge?

💡 Here's a possible solution:

Cost of parts = £79
Hours of labour = 2 × 3 = 6 hours

Amount charged = 79 + 20 × 6
= 79 + 120
= £199

A BIDMAS tells you to complete the multiplication before the addition

I Is the answer appropriate for the situation? For example, an answer of £1.99 would suggest an error has been made

Have a go

1   A school uses the formula below to estimate the number of exercise books students need each year.

   Number of exercise books = Number of students × 2.5
   The school has 820 students.

Q Estimate the number of exercise books the school needs.

2   A water company uses the formula below to estimate the amount of water used in a town each day.

   Water used (litres) = Population × 150
   Whitstable has a population of 30 195 people.

Q Approximately how many litres of water do the people of Whitstable use on a typical day?

3   Nathan is a car salesman.
This formula calculates his wages:

amount earned (£) = hours worked × 6 + number of cars sold × 50

On Monday Nathan starts work at 8.00 am. He sells two cars.
He finishes at 5.00 pm.

(Q) **How much money did Nathan earn on Monday?**
Show all your working.

4   This formula calculates the cost of hiring a carpet cleaning machine:

cost (£) = number of days × 10 + number of extra hours × 2

Olivia hires the carpet cleaning machine from 10.00 am on Monday until
2.00 pm on Thursday.

(Q) **How much will Olivia have to pay?**
Show all your working.

○ We're on the way

In this section you will:
- ◉ use the correct units when substituting values into formulae
- ◉ convert between units of measurement
- ◉ use appropriate units when giving solutions

◯ Take a look: Cooking dinner

The formula below calculates the cooking time for roast beef:

cooking time (min) = weight of joint (pounds) × 25 + 30

Bill wants to roast a 2 kg joint of beef.

Fact

1 kg = 2.2 lb

(Q) **How long in hours and minutes will it take?**

(Q) Here's a possible solution:

A Convert weight from kilograms to pounds

Weight = 2 kg
Weight = 2 × 2.2 = 4.4 pounds
Cooking time (min) = weight of joint (pounds) × 25 + 30

R Identify the correct units for each variable

Cooking time = (4.4 × 25) + 30
= 110 + 30
= 140 minutes
= 2 hours and 20 minutes

ℹ Giving the solution in hours and minutes makes it easier to understand

Have a go

5   The formula below is used to calculate the time it takes to download information from the internet:

$$\text{time (sec)} = \frac{\text{data received (mb)}}{\text{connection speed (mb/sec)}}$$

1 gigabyte (gb) is equal to 1024 megabytes (mb).

Patrick's internet connection has a connection speed of 2 mb/sec.
Patrick wants to download a 3 gb film.

Think First!

Give the solution in hours and minutes.

Q How long will it take?

6   A restaurant pays waiters an hourly wage and gives every waiter a share of all the tips.
The formula below calculates the amount a waiter earns:

amount of pay (£) = hourly wage (£) × hours worked + 0.2 × amount in tips (£)

John earns £6 an hour.
On Monday he works from 6.00 pm until 11.00 pm.
The amount customers give him in tips is shown below:

£3   £4.20   £1   £2.70   £4.50   60p   £1.20

Q Calculate the amount John earns on Monday.
Show all your working.

7   Nick wants to hire a car.
He needs to drive from London to Birmingham then back to London.
He has the choice between two car rental companies:

Rent-a-Car
cost (£) = number of miles × 3 + 100

Car Hire Express
cost (£) = number of days × 30 + number of miles × 2

He will collect the car on Monday.
He will return the car on Thursday.
London to Birmingham is 100 miles.

Q Which company has the best deal for Nick?

Exam ready!

In this section you will:
- find relevant information from charts and tables
- use charts and tables to do calculations
- use charts and tables to communicate results

Take a look: Ordering textbooks

A school orders new maths textbooks for Year 7.
Year 7 has five classes: 7A, 7B, 7C, 7D and 7E.

The table below shows the number of students in each class.

Class	Number of students
7A	28
7B	32
7C	27
7D	23
7E	21

The timetable shows when each class has a maths lesson.

Lesson	Monday	Tuesday	Wednesday	Thursday	Friday
1					
2	7A 7C			7A 7D	7A 7E
3		7A 7B 7E	7E	7C 7E	
4	7B 7D	7D			7C
5					

Each student needs one textbook to use in a maths lesson.
Students don't take textbooks out of the lesson.

Q **a)** How many textbooks does the school need to buy for Year 7?

The school uses the formula below to calculate the cost of new books:

cost (£) = 14 × number of textbooks + 7

Q **b)** Use the formula to calculate the cost of maths books for Year 7.

💡 Here's a possible solution:

a) Maximum number of students needing textbooks = 28 + 32 + 21 = 81

Total number of textbooks = 81

> ℝ Use the timetable to calculate the maximum number of students being taught at any one time. On Tuesday, lesson 3, there are three classes: 7A, 7B and 7E

b) Cost (£) = 14 × number of textbooks + 7
$$= 14 \times 81 + 7$$
$$= 1134 + 7$$
$$= £1141$$

The cost of maths textbooks for Year 7 is £1141.

 Have a go

8 In a mini league, United play Athletico, Wanderers and City.
The results of some of the matches were:

United v Athletico	1 – 2
United v Wanderers	2 – 2
United v City	3 – 0

> **Fact**
> When a team loses a match it earns 0 points.

You can use the formula below to calculate league points for the team:

number of points = number of wins × 3 + number of draws

Q How many points have United got so far?

9 Emma is going to put solar panels on her roof.
You can use this formula to calculate the amount of electrical energy produced by a solar panel:

electrical energy (kilowatthours) =
0.1 × power from sunlight (kilowatts) × hours of sunshine (hours)

A solar panel gets 2000 watts of power from sunlight.

There are 205 hours of sunshine where Emma lives in July.

> **Fact**
> 1 kilowatt = 1000 watts

Q How much electrical energy can Emma expect one solar panel to produce in July?

10 Jane uses this formula to calculate the wages of her employees:

amount to pay (£) = daytime hours × pay rate + unsocial hours × pay rate × 2

The diagram below shows when Michael works for Jane.

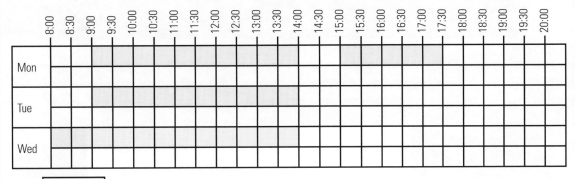

Key: Michael

Daytime hours are between 9.00 am and 5.00 pm.

All other times are unsocial hours.

Michael has a pay rate of £6.

Q Calculate the amount Jane must pay Michael.

Now you can:

- Use correct units
- Convert between different units of measurement
- Use information from tables and charts
- Set out solutions methodically
- Use appropriate units for solutions
- Consider the appropriateness of results

8 Area and perimeter
Know Zone

Area

You need to know that the amount of 2D space a shape takes up is called area.
Units for area are square centimetres (cm²), square metres (m²) and square kilometres (km²).
You can use a square grid to estimate the area of a shape.

Example The map shows the town of Nailsea.
Each square on the red grid represents 1 km².
Use the map to estimate the size of Nailsea.

Answer The approximate area is just over 4 km².

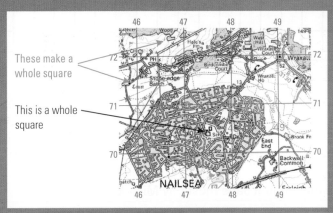

These make a whole square

This is a whole square

You need to know how to find the area of some common shapes using formulae. The formula for calculating the area of a rectangle is:

area of a rectangle = length × width

Example A4 paper is 21 cm wide and 29.7 cm long.
Calculate the area of an A4 piece of paper.

Area = length × width
Area = 29.7 × 21

Answer Area = 623.7 cm²

Perimeter

The distance around the outside of a shape is called the perimeter.
You need to know how to find the perimeter of a shape by adding up the lengths of all the sides.

Example Calculate the perimeter of the field shown in the diagram.

Answer Perimeter = 25 + 25 + 10 + 10 + 30 + 10
= 110 metres

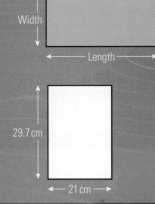
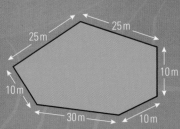

Let's get started

In this section you will:
- sketch a diagram from written information
- calculate area and perimeter to solve a problem
- think about the appropriateness and accuracy of results

Take a look: Football pitch

The Football Association publishes maximum and minimum pitch sizes for different age groups to use.

Age	Max width	Min width	Max length	Min length
Under 11s and Under 12s	50.77	42	82	68.25
Under 13s and Under 14s	56	45.5	91	72.8
Under 15s and Under 16s	64	45.5	82.3	100.6
Under 17s, Under 18s and seniors	90	45.5	90	120

Fact

Under 16s is a team of children who are 15 years old or have just turned 16

Q What is the maximum area of an Under 16s pitch?

Here's a possible solution:

area of a rectangle = length × width

A Use the formula for the area of a rectangle

Maximum area = 64 × 82.3
Maximum area = 5267.2 m^2

 Have a go

1 Here is a plan of Amy's living room.

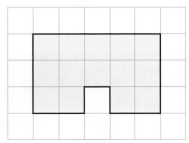

Each square represents 1 m².

Q Find the area of Amy's living room by counting the squares.

2 Here is a satellite image of Gatwick airport on a 1 km² grid.

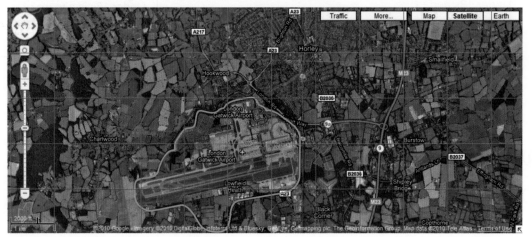

Q Estimate the size of Gatwick airport.

3 John wants to estimate the size of his classroom.
First he measures his stride. One stride is equal to 80 cm.

The classroom is 10 strides wide. It is 18 strides long.

Q Estimate the size of the classroom in square metres.

Think First!

A stride is the length of one step you take when you walk.

We're on the way

In this section you will:
- ◉ calculate building costs for areas
- ◉ think about different solutions to a problem
- ◉ communicate solutions clearly and methodically

🔍 Take a look: Driveway

Jarad has decided to resurface his drive.
The drive is 10 m long and 5 m wide.
A builder charges £9 per square metre for materials.
He charges £100 for labour.

Q Calculate the total cost for resurfacing the drive.

💡 Here's a possible solution:

$$\text{Area of drive} = \text{length} \times \text{width}$$
$$= 10 \times 5$$
$$= 50 \, m^2$$

Cost of material $= 50 \times 9 = £450$
Cost of labour $= £100$

$$\text{Total cost} = \text{Cost of materials} + \text{Cost of labour}$$
$$= 450 + 100$$
$$= £550$$

🎯 Have a go

4 John wants to buy some wood chippings to put on his flower bed.

The local DIY store has the following offer for wood chippings:

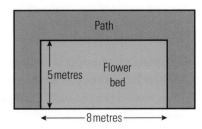

Wood chippings
£2.99
Covers 4 m²

Q How much will it cost John to cover all of the flower bed?

5 Clive has a swimming pool.
He wants a cover for the pool.

The swimming pool is 12 m long and 8 m wide.

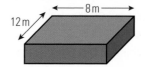

(Q) a) What is the area of the top of Clive's swimming pool?

The material for the cover costs £25 per square metre.

b) How much will the material for the cover cost Clive?

6 Benton wants to make his house warmer. He is going to insulate his loft.
Benton's loft is rectangular in shape.
It has a length of 12 metres and a width of 8 metres.
The table gives information about two different types of loft insulation material.

Type	Size	Cost	Instructions
Roll	4 m × 0.6 m	£8 per roll	Place side by side in strips
Blanket	Covers 8 m²	£30 per sheet	Cut to size needed

(Q) Which type of insulation is the best value for money?

> **Think First!**
>
> When writing your solution, make it clear the calculations that are for
> - roll type
> - blanket type.

7 A regulation table tennis table is 2.74 m long and 1.53 m wide.

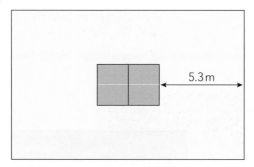

5.3 m

The manager of a sports centre has ordered a table tennis table.
She has worked out where to put the table. Each end of the table has to be 5.3 m away from the wall. Each side of the table has to be 2.7 m away from the wall.

(Q) What is the minimum area of the table tennis room in square metres?

Exam ready!

In this section you will:
- know how to calculate perimeter to solve a problem
- model situations using diagrams
- use diagrams to illustrate the solution to a problem
- use information presented in tables and lists

Take a look: Building a fence

Unal wants to build a fence around the perimeter of his garden.
The garden is shown in the diagram.

The table shows the cost of the fence panels.

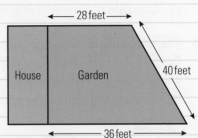

Length	Cost per panel
4 feet	£14
6 feet	£16
8 feet	£22

Unal can cut panels to the size he needs, but he can't
use the pieces left over.

Q How much will Unal need to spend on fence panels?

Here's a possible solution:

The table shows how many panels
Unal needs for each side of the garden.

> **I** Using a table in your solution makes the information easier to read and understand

	4 feet panels		6 feet panels		8 feet panels	
	No.	Cost(£)	No.	Cost(£)	No.	Cost(£)
28 feet	7	98	5	80	4	88
40 feet	10	140	7	112	5	110
36 feet	9	126	6	96	5	110

Total cost = 80 + 110 + 96
= £286

> **A** Think about the cost of using fence panels of different lengths

Have a go

8 Jenny is redesigning her garden.
She has a 2 m × 2 m square of grass.

She places 40 cm × 40 cm paving slabs around the outside of the grass to form a square border.

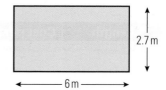

40 cm

2 m

40 cm

40 cm

2 m

Q **a)** How many paving slabs does Jenny need?

Paving slabs are sold in packs of 5.

Think First!

You can draw a sketch to help you

b) What is the minimum number of packs of paving slabs Jenny must buy?

c) Calculate the perimeter around the outside of the border.

9 The diagram opposite shows the wall of a living room.
Damien wants to paint the wall once.
1 litre of paint covers approximately 6 m².
The size and the cost of tins of paint are shown below.

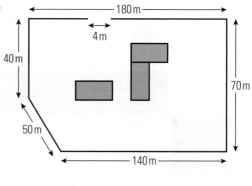

2.7 m

6 m

2 *l*
£13

1 *l*
£6

500 ml
£4.50

Q **a)** How many litres of paint does Damien need?

b) What is the total cost of the paint?

10 A school wants to build a 2 metre high wire fence around its perimeter.
A 4 metre gap is needed for the front gate.
Wire fences are sold in 2 m × 20 m rolls.
They cost £4.99 per square metre.

180 m

4 m

40 m

70 m

50 m

140 m

Q Calculate the total cost of wire fence needed.

11 Ben is redesigning his garden.
He wants to construct a path using pebbles.

The path is 14 m long and 2 m wide.
Pebbles cost £5.99 per 20 kg bag.
40 kg of pepples covers 1 m².

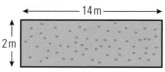

14 m

2 m

Q How much does Ben need to spend on pebbles?

12 James is the manager of a supermarket. He wants to calculate sales per square metre of the supermarket.

The diagram shows selling space of the supermarket.

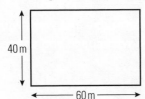

40 m

60 m

Q **a)** Calculate the number of square metres of selling space in the supermarket.

The total sales in the supermarket for the last three months are given in the table.

Store sales (£)	
Month	**Sales (£)**
June	125 880
July	154 320
August	145 650

This formula is used to calculate sales per square metre:

sales per square metre (£/m²)
= total sales (£) ÷ square metre of selling space (m²)

Q **b)** Use your answer to part **a)** to find the sales per square metre of the supermarket for each of the last three months.

Now you can:

- Use information from tables and lists
- Calculate the perimeter of a shape
- Calculate the area of rectangles
- Find missing lengths on rectangles when you know the area
- Present solutions clearly and methodically

In the exam there will be questions that ask you to work with word formulae and work with different units. Some questions will ask you to solve problems to do with space and shape.

Results Plus
Maximise your marks

Question

A health club give advice to customers about how to lose weight.

Cherry wanted to know her weight description, so the health club gave her a formula to calculate her BMI (Body Mass Index).

The formula is only for Cherry to use. It is based on Cherry's height.

$$BMI = \text{weight in kilograms} \times 625 \div 1764$$

The table below shows what the BMI means.

BMI	Weight description
18.5–24.9	Normal
25.0–29.9	Overweight
30.0–39.9	Obese
40 and over	Seriously obese

Cherry has a weight of 97 kg.

What is Cherry's weight description?

Examiner tip

You need to convert Cherry's weight into pounds.
Cherry's weight description can be found from the table.

Student response	Examiner comments
Let's look at a poor answer: $97 \times 625 \div 1764 = 273.7728$	The student has calculated Cherry's BMI incorrectly (they have keyed the signs into their calculator in the wrong order). They have not given a weight description for Cherry.
Let's look at a better answer: $97 \times 625 \div 1764 = 28.0$ Cherry is overweight.	There is an error in the calculation. The student has used the table to find a weight description for Cherry. Marks will still be given for the weight description even though the calculation is wrong.

Student response	Examiner comments
Here is an answer that will gain full marks: $97 \times 625 \div 1764 = 34.4$ *Cherry is obese.*	In this answer, Cherry's BMI has been calculated correctly. The weight description is also correctly given.

ResultsPlus
Exam Question Report

Exam question

A kitchen unit has a width of 600 mm.

Jim has bought the unit to go into a space in his kitchen. The width of the space is 0.5 m.

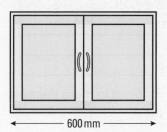

← 600 mm →

Is the space wide enough for the kitchen unit?

Give a reason for your answer.

(3 marks, January 2010

How students answered

66% of students (0–1 mark)

Some students gave a correct decision and compared the unit with the width of the space; however, they didn't write down any evidence of considering a conversion between millimetres and metres.

4% of students (2 marks)

Some students wrote down a conversion, but it wasn't complete.

30% of students (3 marks)

Some students gained full marks by giving a complete conversion, such as metres to millimetres, and a correct decision.

For example, full marks would be gained for the following answer:

No, it would not fit because one metre is 1000 mm and 0.5 m is 500 mm. The unit is 100 mm too big.

Put it into practice

1. Find some word formulae from old exam papers and textbooks. The internet is a good source for finding formulae. Make sure you check with your teacher that the questions are suitable for Level 1.

2. Find the dimensions of some kitchen appliances, for example, fridges or cookers.
   Check the websites of companies that sell these items to find the information.
   Measure the width, depth and height of the space for these appliances in your kitchen (you may need supervision).
   Are some of the appliances too big for the spaces in your kitchen?

ASSESSMENT PRACTICE 2

1 Vic is a plumber. He is working on a job.

Here are the lengths of pipe Vic needs:

Length of pipe	Quantity
2.5 m	2
5 m	2

Vic has these lengths of pipe in his workshop.

Length of pipes (m)	Quantity
1	2
2	2
1.5	1
2.5	1
5	1

Vic can cut the pipes to make them shorter.

He can join the pipes together to make them longer.

Q Vic can use the pipes he's got to do the job. Explain how.

2 Thomas and Georgia visit a family theme park.

They want to watch the sea lion show.

There is a sea lion show three times a day.

The Sea Lion Show Times

Q **a)** The time is 12.35

How long do Thomas and Georgia have to wait for the next sea lion show?

b) What is the time difference between the shows?

3 This formula tells you how tall a boy is likely to be when he grows up.

164 cm   187 cm

> Add the mother's height and the father's height.
> Divide by 2.
> Add 8 cm to the result.

Jenny's height is 164 cm. Sethi's height is 187 cm.
Jenny and Sethi have a baby son.

Q Use the formula to work out the height their son is likely to be when he grows up.

4 A science textbook gives two ways to change celsius (°C) to fahrenheit (°F).

Approximate rule	Exact rule
Double the ˚C temperature	Multiply the ˚C temperature by 1.8
Add 30	Add 32

Q Use the rules to find the temperature of 30˚C in fahrenheit (˚F).

5 A farmer has a rectangular field.

He will put horses on $\frac{1}{3}$ of the area of the field.
The field is 300 m by 90 m.

Q Draw a labelled diagram of the field.
Show in your diagram the area where the horses will be.

6 A manufacturing company makes wire grids. The grids are square.
They have a length of 45 cm.

This diagram shows the grid.
The grid is made from nine small squares of wire.

Q What is the total length of wire needed for the grid?

7 A theme park has four rides, as shown in the diagram.

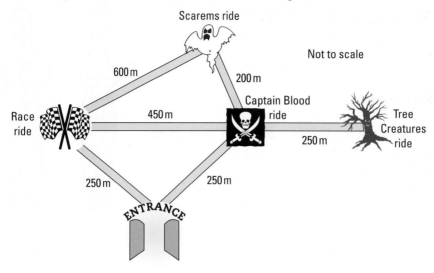

Paths join each of the rides.

Yamina only wants to go on the Race ride and the Scarems ride.
She walks from the entrace to the Race ride. Then she walks to the
Scarems ride. Yamina then walks back to the entrance.

Q **a)** How many metres does Yamina walk?

The owners of theme park are going to put signs to the entrance
on some of the paths.

Entrance 300 m

b) Explain or draw where to put this sign.

8 Tom works in a department store.
He uses this rule to change normal prices to sale prices.

The normal price of a chair is £98.
The normal price of a bed is £149.

Q **a)** Work out the sale price of a chair and the sale price of a bed.

At the end of the sale Tom changes the sale prices back to normal prices.

b) Write down a rule he can be use to change a sale price back
to a normal price.

9 Bella is a college student. She is doing a Textiles course.

She has two textile lectures each Thursday. The lectures start at 1.30pm and 2.45pm. Each lecture lasts 45 minutes.
This Thurday there will be three extra lectures to help students revise.

Revision lectures	
Memory training	1000–1230
Speed reading	1300–1530
Mind mapping	1600–1830

Bella has to go to her two textile lectures. She also wants to go to some of the extra revision lectures. She wants to be in each lecture from start to finish.

Q Which revision lectures can Bella go to?

10 Jane is going to make a 6 m square pattern of tiles. She will use three different sized square tiles.

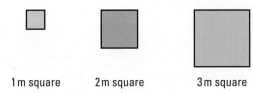

1 m square   2 m square   3 m square

She will use each different sized tile at least twice.

Q Draw a diagram of the 6 m square pattern of tiles.

Know Zone

This chapter is about collecting and recording information, and presenting results.

Data

- You need to be able to collect and record data in a tally table.

- You need to be able to organise **discrete data** in a table
  This table shows the results of a school election.
  The table should have clear headings and be easy to read.

Name of candidate	Number of votes
Andy	13
Cassie	32
Robbie	24

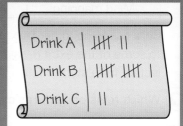

You could use a tally table to record the results of a 'taste test' in a supermarket

The tallies must be put in groups of 5

- You need to be able to choose a sensible scale and label charts, graphs and diagrams.

- You need to be able to represent information in a bar chart, line graph or pictogram.

Pictogram showing the most popular sports in Year 10

Football	🙂 🙂 🙂 🙂 🙂 ✓ for 45 students
Cricket	🙂 🙂 ✓ for 20 students
Hockey	
Athletics	
Basketball	

Give pictograms a title and symbols to represent the number in each category.

Calculate the number in each category by using the key.

Key: 🙂 = 10 students

- You need to be able to describe data that has been presented in graphs, charts and tables.

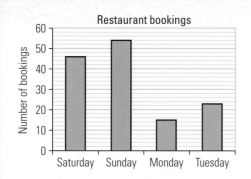

Restaurant bookings

More people are booked in the restaurant on Sunday than on Monday so more staff would be needed

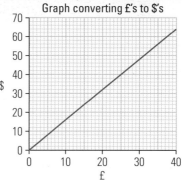

Graph converting £'s to $'s

This graph converts £ to $
Each sub-division on the $ axis is $2
Each sub-division on the £ axis is £1

Label both axes correctly and give the graph a title
Use sensible scales so it is easy to work out each small unlabelled subdivision

Let's get started

In this section you will:
- use a simple key
- make a simple table to collect information
- communicate your answers clearly

Take a look: Tourist information

Vicki works for Durham Tourist Board. She wants to find out the main reasons why people visit Durham. She knows that a lot of people go to see the castle and the cathedral. Some people visit the university.

Q Design a table or chart that Vicki can use to collect the data.

? Here's a possible solution:

Reason	Tally	Frequency
Cathedral and castle		
University		
Visiting family		
Shopping		
Other		

R Some reasons are given in the question. List other reasons why someone would visit a city

R Leave enough space for tallies

R Always have a frequency tally showing the total of the tallies

Have a go

1 On Monday, the temperature in Dubai was 35°C.
Use the key ☆ = 10°C.

Q How would you represent a temperature of 35°C in a pictogram?

2 Becca wants to find out what car colours are most popular.
She decides to count the different coloured cars in the school car park.

Q Design a table to help Becca collect the information she wants.

3 Val has just started working in a continental café with 20 tables.

Customers order tea or coffee, continental breakfast or English breakfast.
Val wants to record customers' orders using a data collection sheet.

(Q) Design a simple chart or table for Val to use.

4 Sunil asks customers in a supermarket which chocolate
they like best, Brand A or Brand B.
Here are his results.

A, A, A, B, B, A, A, B, B, A, A, B, B, A, A, B, A, A, A, A, B, B, B

(Q) Show these results in a tally table.

5 Vicki records the results from the visitors in Durham City in the table below.
Vicki wants to show these results in a pictogram.

Reason	Tally	Frequency
Cathedral and castle	ЖЖ ЖЖ ЖЖ ЖЖ ЖЖ ЖЖ	30
University	ЖЖ ЖЖ ЖЖ ЖЖ ЖЖ ІІ	27
Visiting family	ЖЖ ЖЖ ІІІ	13
Shopping	ЖЖ ЖЖ	10
Other	ЖЖ	5

(Q) Choose a key for the pictogram.

6 Shelley is worried about the large number of vans using the road next to her
school each morning.
Shelley wants to record the types of vehicles using the road between 8:15 am
and 9.00 am.

(Q) Draw a simple tally chart for her to use.

We're on the way

In this section you will:
- decide what kind of graph is appropriate
- construct appropriate scales
- analyse how time can affect data
- give reasons for your choice

Take a look: Car sales

The Classic Car Company needs to show their sales figures for the past five years.

The table shows the information.

Year	2004	2005	2006	2007	2008	2009
Turnover (£s)	240 000	310 000	350 000	290 000	340 000	135 000

Q Draw a suitable chart or graph to show the information in the table.

R You could use a bar chart or a line graph

Q Here are two possible solutions:

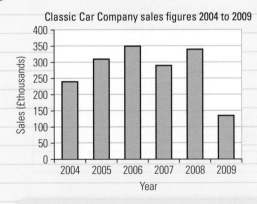

Classic Car Company sales figures 2004 to 2009

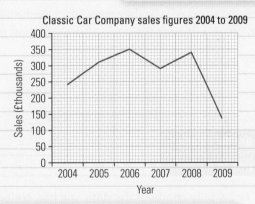

Classic Car Company sales figures 2004 to 2009

R Make sure that both axes are labelled. Find the biggest value that must go on the vertical axis (£350 000). The figures are large, so it is sensible for the scale to go up in £50 000s to a maximum value of £400 000

Have a go

7 Mr Gill is a PE teacher. He asks his Year 10 group the sport they want to choose for fitness work next term.
The table shows the students' choices.

Q Use the table to construct a suitable chart showing the students' choices.

Sport	Numbers
Boxercise	26
Weights	15
Swimming	11
Circuit training	5

8 The table shows the amount of rainfall in Nice.
A travel company wants to show this information in a travel brochure.

Rainfall in mm

Jan	Feb	Mar	Apr	May	Jun	Jul	Aug	Sep	Oct	Nov	Dec
77	74	70	60	47	37	18	31	65	111	117	88

(Q) Draw a suitable graph for the company to use.

9 Acorn computers sell laptop and desktop computers.
The table shows the sales last week.
Dave says that Acorn sold more than twice as many laptops as desktops.

Day	Sales
Monday	D D L L L D
Tuesday	L L L L D D L L
Wednesday	D L
Thursday	L L L L D
Friday	D D L
Saturday	L L L L D L L

(Q) Draw a tally table to show this information.
Use it to help you decide if Dave is correct.

10 Forest workers want to know the different ways visitors use the forest. Visitors can:
- use the cycle tracks
- go for walks
- use the picnic and play areas.

Ellen asks people why they are visiting the forest on one Sunday. She does this at two times, between 9 am and 10 am and then between 4 pm and 5 pm.

Here are Ellen's results.

Key: C = Cycle tracks
W = Walks
P = Picnic and play

9 am to 10 am	C, C, C, P, W, C, W, C, W, C, W, W, W, C, C, C, C, C, W, W, W, W, W
4 pm to 5 pm	P, C, P, P, W, W, W, W, P, P, W, W, P, P, P, P, W

(Q) **a)** Draw tally charts of the data she has collected.

b) Show this information in bar charts.

c) Comment on any difference in use between the two times.

11 Vicram manages a supermarket. One cash till must be open for every 25 customers in the store.

(Q) Use the graph to advise Vicram how many cash tills need to be open between 10 am and 1 pm.

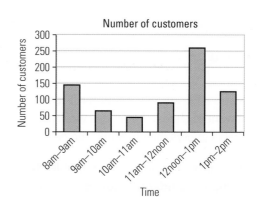

Number of customers

Exam ready!

In this section you will:
- ⊙ decide how to use the information given
- ⊙ use an appropriate plan to solve the problem
- ⊙ make conclusions and communicate your results with reasons

🔍 Take a look: Conversion graphs

The table shows temperatures from 0° Celsius (°C) to 40° Celsius.
It also shows these temperatures in Fahrenheit (°F).

Temp °C	0	20	40
Temp °F	32	68	104

Two friends talk about which temperature is higher, 25°C or 80°F.

Q **a)** Draw a line graph of °F against °C.

b) Use your graph to find which temperature is higher.

💡 Here's a possible solution:

a)

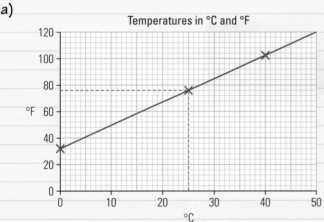

Temperatures in °C and °F

R Use suitable scales for your graph. Here each sub-division on the vertical axis is 4°F

A Plot the points on the graph from the table above. Draw a straight line through these points

b) 25°C is 76°F

So a temperature of 80°F is about 4°F higher than 25°C.

A Draw a vertical line from 25°C. Where this line meets the graph line, draw a horizontal line to the °F axis. Read off the Fahrenheit temperature

Have a go

12 The table shows the numbers of red squirrels in Notherrn Scotland from
    2000 to 2010.

Year	2000	2002	2004	2006	2008	2010
Red squirrel numbers	456	345	276	302	340	498

Q a) Draw a suitable bar chart for the data.
    b) Comment on the number of red squirrels between 2000 and 2010.

13 Speeds in Europe are measured in kilometres per hour (km/h).
    In the UK, speeds are in miles per hour (mph).
    The table below shows the same speeds in mph and in km/h.

Think First!

You should have mph on the
x axis from 0 to 100 and km/h
on the y axis from 0 to 160.

Speed in mph	0	50	100
Speed in km/h	0	80	160

Q a) Draw a graph of km/h against mph.
    b) What speed is 100 km/h when measured in mph?

14 The table shows the stopping distances in metres for a car travelling at a given
    speed in mph in dry weather.

In wet weather, stopping distances are multiplied by 1.5

The Kill your Speed campaign group wants to show how the stopping distances
change for different speeds.

Q a) Work out the stopping distance in wet weather for the different speeds.
    b) Draw a suitable graph or chart the campaign group can use to show the
        stopping distances for dry and wet weather.

Speed	Thinking distance	Breaking distance	Overall stopping distance in dry weather
20 mph	6 m	6 m	**12 m**
30 mph	9 m	14 m	**23 m**
40 mph	12 m	24 m	**36 m**
50 mph	15 m	38 m	**53 m**
60 mph	18 m	55 m	**73 m**
70 mph	21 m	75 m	**96 m**

15 John thinks some new tyres will last for 35 000 km. He asks eight drivers to test the tyres for two years. At the end of the two years, John measures how much tread is left on the tyres.

The law says a driver must buy a new tyre if there is 1.6 mm or less tread left on a tyre.

The results are below.

Distance covered in km	5 000	8000	14 000	17 000	21 000	25 000	28 000	30 000
Depth of tread in mm	7.0	6.5	5.4	4.9	4.2	3.5	3.0	2.6

Q **a)** Draw a graph of distance covered to depth of trend.

b) Do you think the tyres will last for the full 35 000 km?

Now you can:

- Construct a table to collect data
- Use a tally table
- Construct your own graphs and charts
- Present a graph to show information for a purpose
- Use graphs and charts to solve problems

10 Use and interpret data
Know Zone

You will see examples of tables, graphs and charts in newspapers, magazines or company brochures.

Tables, graphs, charts

- You need to know that the title, labels and key provide useful information
  Look at the title and both axes to find out what the graph is showing.

- You should know how to read a scale on an axis.
  When reading a scale it is important to work out what each small sub-division is worth.

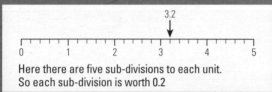

Here there are five sub-divisions to each unit.
So each sub-division is worth 0.2

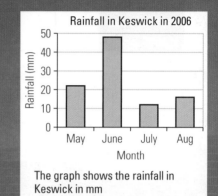

The graph shows the rainfall in Keswick in mm
The graph shows the rainfall in the summer months in 2006

- You need to be able to find information from tables (e.g. timetables or price lists), charts (e.g. pictograms, pie charts, bar charts, line graphs) and diagrams (e.g. maps, workshop drawings or plans)

  When finding information from a table, find the correct row and column, then move along and down to find the figure you want.

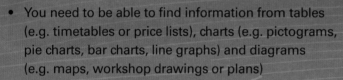

Accommodation	Sunset Holiday Village			Child Prices for Sunset	
Board basis	All Inclusive				
Holiday number	HS1649				
Price based on	Twin Adult			1st Child	2nd Child
No of nights	7	14	Ex Day	Any duration	
01 May–14 May	459	629	28	364	276
15 May–21 May	474	689	28	379	284
22 May–28 May	544	729	28	449	359
29 May–09 Jul	539	739	33	424	314
10 Jul–23 Jul	639	964	46	489	334
24 Jul–10 Aug	789	1144	54	609	429
11 Aug–20 Aug	759	1149	54	579	399

The cost of a holiday to the Sunset Holiday Village for a child leaving on the 10th of July is £489

A line graph can show trends in sales over a period of time.

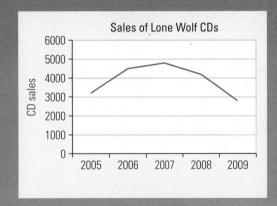

Sales of Lone Wolf CDs

From 2005 to 2007 the sales of Lone Wolf CDs rose. From 2007 to 2009 the sales of the CDs fell. A line graph can show highest points and lowest points.

The sales of the CD were at their highest level of 4800 in 2007.

Mean, range

We hear the word 'average' used all the time.

Examples The temperature is around average for the time of year
The average height for a UK man is 5 feet 9 inches
The average UK house price went down by 2.8% last year

- You need to know how to calculate the mean.
- You need to know how to calculate range.

The type of average used in the above examples is called the **mean**.
To find the mean score you add up all the values and then divide by the number of values.
To find the range you subtract the smallest value from the largest value.

Example A cricketer scored 40, 81, 31, 0, 6, 7 and 12 in a cricket series in South Africa.
The cricketer has a mean score of $177 \div 7 = 25.3$.
The cricketer has a range of $81 - 0 = 81$

- You need to know:
  - that mean is a single value that represents the data
  - that range is a value that measures the spread of the data.

Example A teacher is comparing the Maths and English results for his class. All the results can be represented by the mean and the range.

	Mean	Range
Maths results	74.9	43
English results	65.8	20

Maths results were higher (on average) than the English results. Maths results were more spread out than the English results

Let's get started

In this section you will:
- ◉ choose the correct values to find the range
- ◉ understand simple tables and dials
- ◉ find a mean and range from a set of data
- ◉ make conclusions from calculations involving the mean or range

Take a look: The charity collection

A school collected money for charity.
The table shows the amount of money collected each day.

Monday	Tuesday	Wednesday	Thursday	Friday
£32.45	£15.60	£13.90	£9.60	£24.40

Q Calculate the mean amount collected per day.

💡 Here's a possible solution:

£32.45 + £15.60 + £13.90 + £9.60 + £24.40 = £95.95 ●——— **A** Add up the amounts collected each day

£95.95 ÷ 5 = £19.19 ●——— **A** Then divide by the number of days

Take a look: The range in Spain

One week in August, the midday temperatures in Spain were 30°C, 33°C, 31°C, 34°C, 32°C, 32°C and 28°C.

Q Calculate the range in temperatures.

💡 Here's a possible solution:

Range = 34 – 28 = 6°C

A Find the largest and smallest temperatures

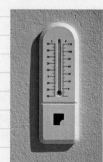

Have a go

1 Janveer records the following prices of bread in a supermarket.
   The prices are: 79p, 84p, 96p, 83p and 90p.

(Q) Calculate the range in these prices.

2 The table shows the cost of a week's holiday to a holiday village
   at different times of the year.
   One adult and two children are going on a two-week holiday,
   starting on 25th July.

Date of departure	Adult price	Child price
7th May to 24th July	£149	£76
25th July to 31st August	£238	£125
1st September to 20th October	£125	£75

(Q) Calculate the price of the holiday.

3 A call centre worker makes ten phone calls in 40 minutes.
   Her target is to talk for an average of 5 minutes or less per phone call.

(Q) Is she meeting her target?

4 A garage sells four types of tyre.
   The sign shows the cost of one tyre of each type.

(Q) What is the average cost of one tyre?

Make	Price
Goodright	£67
Freestone	£56
Unitread	£48
Brimoulds	£40

5 Keith needs to sell an average of £200 worth of
   goods each day to make a profit.
   The table shows the value of the goods Keith sold from Saturday
   to Monday.

	Saturday	Sunday	Monday
Values	£245.60	£312.90	£140

(Q) Use the information in the table to find out if Keith made a profit or not.

6 Rosie wants to go to a holiday resort with a high temperature.
   The table shows information about the temperatures at
   three resorts.

Think First!
You need to compare both
the mean and the range.

Resort	Mean midday temperature in °C	Range in midday temperatures °C
Graffe	34.1	12
Miramar	34.1	2
Tinmouth	22.1	3

(Q) Use the information provided to advise Rosie which resort
to choose (Graffe, Miramar or Tinmouth).

7 The graph shows the number of copies of a magazine sold at a supermarket one week.

Q How many copies of the magazine does the supermarket sell that week?

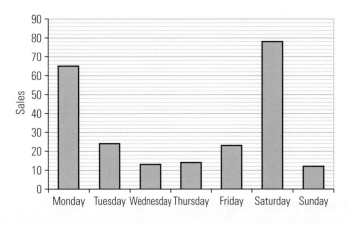

We're on the way

In this section you will:

- choose appropriate measures to compare two sets of data
- calculate the means and ranges of two sets of data
- compare means and ranges; communicate the result of your calculations

Take a look: The golf team

David and Peter play golf each week.
They both want to play for their golf club team.

The table shows their scores for the past eight weeks.

	Week 1	Week 2	Week 3	Week 4	Week 5	Week 6	Week 7	Week 8
David scores	74	76	77	74	79	74	80	75
Peter's Scores	76	77	76	76	78	76	76	77

Q Who you would choose to play in the golf club team, David or Peter?

Here's a possible solution:

Mean score for David = (74 + 76 + 77 + 74 + 79 + 74
    + 80 + 75) ÷ 8 = 76.125

R The mean score will give a single value that represents their scores but it is always a good idea to look at the range

Range for David = 80 − 74 = 6

Mean score for Peter = (76 + 77 + 76 + 76 + 78 + 76
    + 76 + 77) ÷ 8 = 76.5

A Add up the scores for each person and divide by 8 to find the mean

Range for Peter = 78 − 76 = 2

I would choose Peter because he is more reliable.

I The mean scores are almost the same but the range is different

I Peter's score has a range of 2. David's score has a range of 6. Peter's score has a smaller range. So Peter is more consistent

Have a go

8 The table gives the number of days of rain in Cleverley and Birn during the summer.

	May	June	July	Aug	Sept
Cleverley	4	3	2	2	5
Birn	4	0	2	4	8

Q Where does it rain most often, Cleverley or Birn?

Think First!
Use both the mean and the range.

9 Mr Cook is comparing his students' results in their last two Maths tests.

Test 1	56%, 76%, 25%, 78%, 45%, 67%, 81%, 65%, 32%, 67%
Test 2	45%, 86%, 23%, 62%, 40%, 69%, 89%, 62%, 23%, 65%

Q What do the results show about how well the students did in the two tests?

Think First!
Use the mean to calculate the average result for each test. Use the range to find the spread of marks.

10 Dennis and Mary are going on a cruise to celebrate Dennis's retirement.
They are comparing two cruise ships, *The Queen Beatrice* and *The Queen Victoria*.
The pie charts show what people have said about the cruise on each ship.

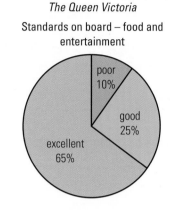

Q Use the information in the pie charts to advise Dennis and Mary on the ship to choose for their cruise.
Give reasons for your choice.

11 John works at a car showroom.
He gets a bonus if he sells an average of three cars a week for a year.
The bar graph shows his sales for the first six months.

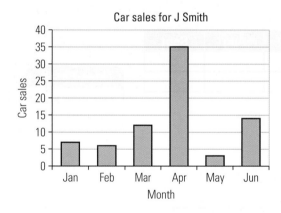

Q **a)** How many cars does John need to sell in a year to get a bonus?

b) Is John going to get a bonus at the end of the year?

Exam ready!

In this section you will:

- decide on the information you will use in the question
- compare means and ranges
- make conclusions from your calculations and communicate your results

Take a look: Maternity unit

Morag and Liskelia work in the maternity unit in a Scottish hospital. The table shows information about the babies born in the summer of the last two years.

Q

a) Liskelia says that babies born in the summer are more likely to be male. Use the information to decide if you agree with her.

b) Morag says that male babies are heavier than female babies. Compare the weights of the male and female babies to see if Morag is correct.

2008			2009		
Sex	**Month born**	**Weight (kg)**	**Sex**	**Month born**	**Weight (kg)**
M	May	3.5	M	June	4.1
F	May	3.7	M	June	4.8
F	May	3.2	F	June	4.3
M	June	3.9	F	July	4.2
M	July	4.5	M	July	5.0
M	July	4.9	M	Aug	3.9
F	Aug	4.2	M	Aug	3.6
			M	Aug	4.2

Here's a possible solution:

a) Here is a tally to find the sex of the babies born in these months.

Summer month	Male babies	Female babies
May				||
June	|||		
July	|||		
August	|||		
Total	10	5

ℹ There are twice as many male babies as female babies. Do you think that this finding will be representative of the UK as a whole?

Liskelia is correct. The results in the table show babies born in the summer are more likely to be male than female.

b) There are 10 male babies with a mean weight of 42.4 ÷ 10 = 4.24 kg.
There are 5 female babies with a mean weight of 19.6 ÷ 5 = 3.92 kg.
Morag is correct. The results in the table show that male babies are heavier
by 0.32 kg.

Have a go

12 Freda has 18 tomato plants.
She is going to test three different fertilisers on the plants to find out which fertiliser is best.

Freda gives all the plants the same amount of water.
She keeps all the plants at the same temperature.
Freda puts • her usual fertiliser on 6 of the plants
          • seaweed fertiliser on 6 of the plants
          • chemical fertiliser on 6 of the plants

Freda records the total weight of tomatoes each plant produces.
Her results are shown in the table.

Usual fertiliser	
	Yield (kg)
Plant 1	2.3
Plant 2	1.8
Plant 3	3.1
Plant 4	2.8
Plant 5	3.6
Plant 6	2.8

Seaweed fertiliser	
	Yield (kg)
Plant 1	2.8
Plant 2	3.0
Plant 3	2.2
Plant 4	Died
Plant 5	3.0
Plant 6	3.2

Chemical fertiliser	
	Yield (kg)
Plant 1	2.7
Plant 2	3.1
Plant 3	2.8
Plant 4	3.0
Plant 5	1.9
Plant 6	2.1

Q Use the information in the table to find out the fertiliser that gives the greatest
weight of tomatoes.

13 Peter, Val and their daughter Lindsey want to go on holiday to the Hotel Sol. They have saved up a total of £1250. They like to go on holiday when it is as hot as possible.

Hotel Sol	Adult price		Child price	
Departures between	7 days	10 days	7 days	10 days
1st May to 15th June	£ 356	£512	£150	£150
16th June to 22nd July	£402	£586	£150	£150
23rd July to 1st September	£520	£690	£250	£250
2nd September to 12th October	£380	£545	£150	£150

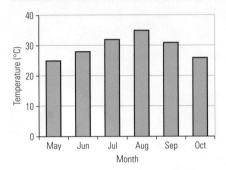

Q **a)** Use the information above to advise them when they should have their holiday.

b) How long can they go on holiday for?

14 A golf club needs to raise £30 000

Option 1: each member pays £50

Option 2: raise the annual subscriptions by 10%.

	Number of members	Annual subscription
Full member	340	£700
Student or OAP member	190	£340
Social (non playing) member	80	£100

Q **a)** Work out the cost of an annual subscription for a full member if annual subscriptions are raised by 10%.

b) Use the information in the table to work out if the golf club will get the £30 000 it needs.
Which is the best option for the golf club?

Now you can:

- Find the mean and range of a set of data
- Understand that the mean is affected by low (or very low) values
- Understand that the range measures the spread of the results
- Extract information from graphs and tables
- Read and use a scale from an axis

11 Likelihood of events
Know Zone

We often talk about how likely an event is to happen.

These are all questions we ask in everyday life:
- What are the **chances** of winning the lottery?
- What is the **likelihood** of rain today?
- What are the odds of Manchester United winning the Champion's League?

Events

You need to know some key facts to help you decide about likelihood.

The result of an event is called the outcome.

Some events are more likely to happen than others.

Events are random if they are unplanned and not affected by each other.

Events can be:
- certain
- likely
- unlikely
- impossible

Likelihood

You can compare the likelihood of events happening and put them in order.

Example Put these events in order of likelihood.
Start with the least likely to happen.
A You will have flu in your lifetime
B Water will freeze at 100°C.
C You will see a flying saucer.
D A baby will be a boy or a girl.

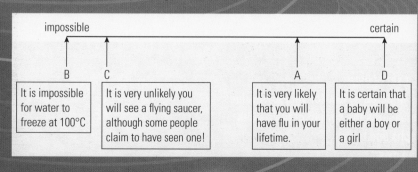

⊙ Let's get started

Let's get started

In this section you will:
- ◉ decide if events are certain, impossible, likely or unlikely

🔍 Take a look: Deciding the likelihood of events

Think about these events:
- You will win the National Lottery.
- The sun will rise tomorrow.
- When you throw an ordinary dice it will land on a 7.
- You will see someone you know today.

(Q) Are these events certain, impossible, likely or unlikely?

(💡) Here's a possible solution:

It is **very unlikely** that you will win the National Lottery because there are so many different combinations of numbers that can be chosen.
It is **certain** that the sun will rise tomorrow.
It is **impossible** to throw a 7 with an ordinary dice because a dice only has six faces with the numbers 1 to 6.
It is **likely** you will see someone you know today.

⊙ Have a go

1 This spinner is used in a game.

(Q) Which colour is the spinner **least likely** to land on? Why?

2 Last week Gavin had an accident and broke his arm.

(Q) What is the likelihood of him breaking his other arm this week?

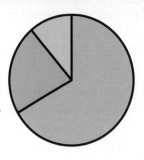

3 On his way to work today, Wes found £10 in the street.

(Q) What is the likelihood that he will find £10 in the street tomorrow?

4 Think about these events:
- A It will be light all night in England.
- B When you throw an ordinary coin it will land on heads.
- C If you live a healthy life you will reach the age of 60.
- D When you throw a dice it will land on a 6.

(Q) Put these events on a scale. Start with the least likely to happen.

We're on the way

In this section you will:
- find possible outcomes
- know that some events can happen in more than one way
- know when events are equally likely

Take a look: Playing cards

Annamaria has a pack of 52 cards. She picks a card at random.

(Q) **a)** **i)** How many red cards are in a pack?

ii) Is Annamaria more likely to pick a red card than a black card?

b) Annamaria picks a 10 from the pack.
List all the cards she could have picked.

Here's a possible solution:

a) **i)** The total number of red cards is
13 hearts + 13 diamonds = 26.

ii) No. There are 26 red cards and
26 black cards in the pack, so
Annamaria is equally likely to pick a
red card as to pick a black card.

b) There are four 10s in a pack of cards:
10 of hearts, 10 of diamonds, 10 of
clubs and 10 of spades.

Fact

A pack of cards has 52
cards. The cards are: 2, 3, 4,
5, 6, 7, 8, 9, 10, Jack, Queen,
King and Ace.
There are four suits. Each
suit has 13 cards.
Hearts and diamonds
are red cards. Clubs and
spades are black cards.

i) You know Annamaria
picked a 10, so there are only
four possible outcomes

Have a go

5 Meryll throws two different coloured dice.
She adds the numbers together to get her score.

(Q) Meryll scores 7. List all the possible ways she can score a 7.

Think First!

The dice are different
colours. What difference
does this make?

6 Joe wants to find out how students travel to college.
He does a survey.

(Q) List the possible methods of travelling to college.
Use your experience to describe the likelihood of each method of travelling.

7 A couple want to have three children.
It is equally likely that each child will be a boy or a girl.

(Q) List the possible outcomes for the three children.

Think First!

Think about what order the
children are born in.

Exam ready!

In this section you will:
- use real data to decide on the likelihood of events

🔍 Take a look: Crime in Manchester

Jemima wants to compare the number of crimes in Manchester in 2008 and 2009.

The table shows the number of crimes reported in Manchester in the last three months of 2008 and 2009.

	Oct	Nov	Dec
2008	6599	6602	6117
2009	6039	6220	5749

Jemima thinks that people living in Manchester were less likely to be affected by crime in the last three months of 2009 than in the last three months of 2008.

Q Do you agree with Jemima?

💡 Here's a possible solution:

The total number of crimes reported in Manchester in 2008 was 19 318.

R You need to look at all the months shown in the table

The total number of crimes reported in Manchester in 2009 was 18 008.

There were 19 318 − 18 008 = 1310 fewer crimes reported in Manchester in 2009.

I agree with Jemima. There were fewer crimes in 2009, so people living in Manchester were less likely to be affected by crime in 2009.

I Write your answer clearly. Remember to give your reasons

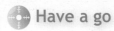 Have a go

8 Jessica and Joe are are going to get married in Sheffield. They want to have an outdoor wedding in June, July or August.

The first table shows the number of days it did **not** rain in Sheffield for June, July and August for 2007–09. The second table shows the total rainfall for these months.

Number of days with no rain			
	June	**July**	**Aug**
2007	12	8	22
2008	12	13	11
2009	19	5	12

Total rainfall (mm)			
	June	**July**	**Aug**
2007	245	125	27
2008	54	97	110
2009	122	141	49

Think First!

What statistical methods can you use to help you make a decision?

Q Use the information in the tables to advise Jessica and Joe in which month they should have their wedding.

9 The table shows the number of car drivers in Great Britain who were injured in accidents in 2006.

Age in years	under 19	20–29	30–39	40–49	50–59	60–69	70–79	80 and over
Number injured	11 656	30 548	22 944	19 596	11 891	6257	3370	1624

Farhan says this shows that the safest car drivers are aged 80 and over.

Q Farhan does not have all the information he needs to decide which age group is the safest.
What other information does he need?

ResultsPlus
Exam Tip

Set out your working clearly, so that the examiner can see how you decided on your answer. Give reasons for your decision.

Now you can:

- Decide if events are certain, impossible, likely or unlikely
- Find all the possible outcomes of an event
- Know that some events can happen in more than one way
- Know when events are equally likely
- Use real data to decide on the likelihood of events

Robbie and Chardonnay want to buy a holiday apartment.
They want to buy an apartment in the area that has the warmest and driest weather.
Robbie and Chardonnay find some weather statistics for two regions they like.

Costa Blanca

Months May to September	M	J	J	A	S
Average maximum temperature °C	24	27	31	32	28
Average monthly rainfall mm	32	17	6	8	48

Playa Des Las Americas

Months May to September	M	J	J	A	S
Average maximum temperature °C	23	27	29	30	27
Average monthly rainfall mm	6	0	0	0	2

Where should Robbie and Chardonnay buy an apartment, Costa Blanca or Playa Des Las Americas?

Use calculations to support your decisions.

Examiner tip

Decide what calculations you are going to do.
Then write down decisions and comparisons based upon your calculations.

Student response	Examiner comments
Let's look at a poor answer: I would recommend that Playa is the best place to stay.	This answer makes a decision but where are the calculations? The answer does not say how the student has made this decision and why they chose Playa. This response has not answered the question and would not gain any marks.
Let's look at an answer that would gain some marks: Playa Des Las Americas is the coldest place in this period of time. Costa Blanca is the warmest. Playa has the least rainfall and Costa Blanca has the most.	This answer has written down some comparisons between temperature and rainfall. However, the comparisons are not based upon calculations and no decision is made.

Student response	Examiner comments
Let's look at an answer that would gain full marks: Mean temperature: Costa Blanca $(24 + 27 + 31 + 32 + 28) \div 5 = 28.4°C$ Playa Des Las Americas $(23 + 27 + 29 + 30 + 27) \div 5 = 27.2°C$ Mean rainfall: Costa Blanca $(32 + 17 + 6 + 8 + 48) \div 5 = 22.2$ mm Playa Des Las Americas $(6 + 0 + 0 + 0 + 2) \div 5 = 1.6$ mm There is only a small difference between temperatures. Costa Blanca has more rainfall. They should choose Playa Des Las Americas.	This student has made a decision based on comparisons. The calculations the student did to make the comparisons are correct. The choice of calculations is the student's. This student has used means. The comparisons made and the final decision are the student's. Students can also get full marks by making different decisions. The student could have said, 'Rainfall is too low in Playa Des Las Americas'.

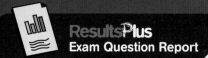

ResultsPlus
Exam Question Report

Exam question

This is a record of the money Kevin is paid during January.

Date	Amount	Date	Amount
05/01/10	£1560	19/01/10	£4740
08/01/10	£5640	20/01/10	£1590
11/01/10	£2480	25/01/10	£3410
14/01/10	£3280	28/01/10	£2300
18/01/10	£2160	29/01/10	£3270

Draw a graph or chart to represent this information for January.   (4 marks, January 2010)

How students answered

■ **39% of students (0–1 marks)**
Some students got a mark for displaying an incomplete graph or chart.

● **26% of students (2 marks)**
Some students had the wrong scales on the axes, but they got marks for a completed graph or chart that was correctly labelled.

▲ **35% of students (3–4 marks)**
Some students found it hard to draw consistent scales for amount or date. Plotting points was generally more successful..

Put it into practice

1   Find some data of your own from a variety of sources, for example, newspapers or the internet.
    Do some different statistical calculations, draw graphs and describe your findings.

2   Look at a number of sports league tables (for example, football, hockey, rugby, netball).
    What are the similarities and differences between them? Do they record the same information?

ASSESSMENT PRACTICE 3

1 Joseph asks people this question in a survey:
How good is your council?

People had five choices for their answers.

Very Good	Good	Satisfactory	Poor	Very Poor

Joseph is going to show his answers in a pie chart.
He says, 'There are five choices, so I need to draw five equal parts in the pie chart.'

(Q) Is Joseph right?
Give reasons for your answer.

2 Angela collected data about film lengths and the age classification of each film as part of a school project.

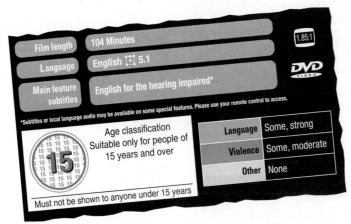

Here are Angela's results.

Age classification	Length (min)
12	121
PG	89
U	97
PG	122
U	76
U	83
PG	90
U	103
12	86
15	132
15	104

(Q) **a)** Put the films in order of length. Start with the shortest length.

b) Is there evidence that the age classification increases as the film length increases? Use calculations to support your answer.

3 A council collected data on the number of people having to pay a fine in 2010.

Age group (years)	2010
17–29	62
30–39	40
40–49	28
50–59	18
60+	12

Q Show the data in a graph or chart.

4 A car park attendant collected information about the tickets people bought from the ticket machines.

Time Period	Cost	Frequency
Up to 2 Hours	£3.00	8
Up to 4 Hours	£4.60	5
Up to 6 Hours	£7.20	4
Up to12 Hours	£8.00	3
Up to 24 Hours	£9.00	1

Q Which time period made the most money for the owners of the car park?

5 A hotel manager collected information on the types of rooms people stayed in over four days.

Rooms	Thu	Fri	Sat	Sun
Single	3	8	3	3
Double	27	33	35	24
Twin	20	24	24	19
Deluxe	6	8	7	3
Suite	3	5	6	1

Q **a)** On which day was the largest number of rooms used?

Dan is writing a report for the hotel manager. He starts to draw a bar chart to show the information from the table for Sunday.

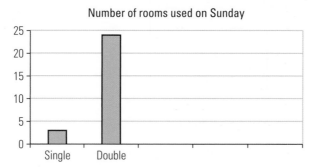

Number of rooms used on Sunday

b) Copy and complete the bar chart for Sunday.
Show all the information needed.

6 Jamilia has 12 scarves.

This table shows some information about the scarves.

Scarf design	Number of scarves
Stripe pattern	3
All white	1
All grey	4
Spot pattern	4

Jamilia takes a scarf at random.

Q Is she most likely to take a scarf with a pattern?
Explain your reasoning.

7 Jeremy is planning a meal out to celebrate Christmas.
He needs to find out what each of his friends would like to eat at the meal.
The meal will consist of 3 courses.

Starter	Main Course	Dessert
Soup	Turkey	Christmas Pudding
Melon	Beef	Cheesecake
	Vegetable curry	

Q Design a data collection sheet that Jeremy can use to record the choice
of each person coming to the meal.

8 A school has five year groups.

The school had a sponsored walk to raise money for charity.

The deputy head drew this pie chart to show the total amount of
money each year group raised for charity.

Amount of money raised

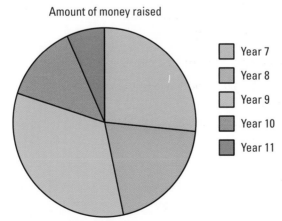

- Year 7
- Year 8
- Year 9
- Year 10
- Year 11

Q **a)** Which year group raised the least amount of money?

The Year 9 students raised £200.

b) Estimate how much each of the other year groups raised.
Then draw a bar chart to show the information.

9 An Indian restaurant wants to find out the ages of its customers buying take-away meals.
It designs a questionnaire.

One question is:

How old are you (in years)?
Tick the correct box

☐   ☐   ☐   ☐   ☐

20 or younger   20 to 30   30 to 40   40 to 50   50 to 60

Q **a)** The restaurant needs to correct this question.
What part of this question does it need to correct?

Here is a different question the restaurant asks:
How much do you usually spend each week on take-away food?

b) Write down a suitable way to record people's answers.

EXAM-STYLE PRACTICE

1 The table shows the weights of fuel a plane uses for three different stages of a flight.

Weight of fuel used (per hour)	
18 000 kg	Climbing
12 000 kg	Cruising
10 000 kg	Descending

A flight takes a total of three hours. The plane takes 15 minutes to climb to the cruising height. It takes 30 minutes to descend.

Q **a)** How much fuel does the plane use to descend?

b) Calculate the weight of fuel the plane uses for the whole flight.

2 Ole is working on a project about world population.

The table shows the population of six countries in 1950.
It also shows the estimated populations of these countries in 2050.
(The populations are rounded to the nearest million.)

Country	Population (in millions) 1950	Country	Estimated population (in millions) 2050
China	563	China	1424
India	370	India	1808
USA	152	USA	420
Russia	102	Russia	109
Pakistan	40	Pakistan	295
UK	50	UK	64

Q **a)** What is the estimated population of India for 2050?

In 1950, 1 person lived in Russia for about every 3.7 people living in India.

b) In 2050 what is the estimated number of people living in India for 1 person living in Russia?

c) Ole thinks that from 1950 to 2050 the population of Russia will go up by 100%.
Ole is wrong. Explain why.

3 Gunnar is a tiler. He can tile 6–8 m² of floor per day. Gunnar estimates that he takes half a day to buy materials and to clear up when he finishes.

Gunnar is going to tile two rooms:

    Room 1: A square room with sides 3 m.
    Room 2: A rectangular room with sides 4 m by 5 m.

Gunnar must estimate the number of days it will take him to tile the two rooms.

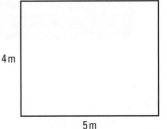

Q Work out an estimate for the number of days.

4 The table shows weight ranges (kg) for different-sized cats.

	Cat Size								
	Short cat			Medium cat			Tall cat		
	Normal	Overweight	Obese	Normal	Overweight	Obese	Normal	Overweight	Obese
Weight (kg)	4.0	4.4	4.8	5.0	5.5	6.0	6.0	6.6	7.2

Stuart has two cats named Stripes and Goth.
Stripes is a short cat.
Goth is a tall cat.
Stripes weighs 4.5 kg and Goth weighs 6 kg.

Q **a)** **Use the table to decide whether Stripes and Goth are normal, overweight or obese.**

Stuart can feed Goth on tinned cat food or complete cat food.

Tinned cat food
A pack of 10 tins costs £3.00
Recommended daily use for cat weighing 6 kg
2 cans per day

Complete cat food

Pack weight	Price
500 g	£2.24

Recommended daily use for cat weighing 6 kg
100 g per day

Stuart wants to buy the cheapest food for Goth.

b) **Which cat food should he buy?**

5 Some runners are training for a marathon.
The table shows some information about four types of runner and the time they would expect to take to complete different length training runs.

		Distance in km		
		6	8	10
Type of runner	Leisure	50 min	70 min	90 min
	Club	42 min	63 min	75 min
	Contender	38 min	42 min	47 min
	Elite	33 min	39 min	43 min

Jessica is a club runner. Harriet is an elite runner.
They both start an 8 km training run at 0730

Q **a)** **At what time would you expect each woman to finish the training run?**

Diana wants to train for the marathon.
She completes some 6 km training runs and records her times
(to the nearest minute).

Here are her results:

53 min, 48 min, 46 min, 47 min, 45 min, 47 min, 49 min,
46 min, 47 min

Diana uses her mean time to decide what type of runner she is.

b) Find the mean of Diana's times.

c) What type of runner is Diana?

6   A council asks a printing company to print some booklets.
    Here are the prices for printing the booklets:

Price	£450	£490	£510	£530	£540
Copies	up to 5000	up to 6000	up to 7000	up to 8000	up to 10 000

The council wants 7500 booklets.

(Q) a) How much will the council pay the printing company?

The printers use this formula for printing more than 10 000 copies:

Total cost (pounds) = Price for 10 000 copies + Number of extra copies ÷ 12 000

A charity asks the company to print 80 000 booklets.

b) What price will the charity pay to the printers?

7   The table shows a company's sales figures for 2009.

Month	Sale figures for 2009 (£)	Month	Sale figures for 2009 (£)
January	20 000	July	27 000
February	26 000	August	23 000
March	24 000	September	26 000
April	18 000	October	34 000
May	25 000	November	31 000
June	30 000	December	25 000

(Q) a) Draw a line graph to show this information.

b) Are the sales figures increasing?
    Give a reason for your answer.

8 The table shows the prices a cinema charges for two different types of seat and for two different types of film.

	Standard seat		Premier seat	
	2D film	3D film	2D film	3D film
Adult	£7.50	£9.55	£8.50	£10.55
Child	£5.50	£7.30	£6.50	£8.30
Family (1 adult + 3 children)	£23.00	£30.00	£27.00	£34.00
Family (2 adults + 2 children)	£23.00	£30.00	£27.00	£34.00

A child goes to see a 3D film.
The child sits in a premier seat.

Q **a)** How much does the child pay?

Debbie and her children, Jake and Sonia, are going to the cinema.
She has a budget of £25. They will sit together.

b) What seats and what types of film can Debbie choose?

Debbie and her two children want to see two films: *Nanny McPhee* and *How to Train your Dragon*. They want to see the films on Saturday.

This programme shows the start times of the two films and the length of each film.

Nanny McPhee
109 min

| Fri–Tue | 10:40 | 11:40 | 13:15 | 14:30 |
| | 15:50 | 17:15 | 20:00 | |

| Wed–Thur | 13.15 | 14.30 | 15.30 | 16.40 |
| | 20.15 | | | |

How to Train your Dragon
98 min

| Wed–Sun | 11:00 | 13:30 | 16:00 | 18:20 | 20:50 |
| Mon–Tue | 10:30 | 13:30 | 15:30 | | |

Debbie and her children will see each film from start to finish.
They can get to the cinema any time after 12 noon. They must leave the cinema by 8 pm.

c) Write down one way that Debbie and her two children can see the two films.

Level 1: Let's get started answers

Chapter 1: Number

1 Joe needs 10 × 142 = 1420 bricks.

2 Four thousand three hundred and twenty-five pounds.

3 £111 500

4 **a**

Team	Ground	Capacity
Hull	Kingston Communication	25 504
Wolverhampton	Molineux	29 400
Everton	Goodison Park	40 569
Chelsea	Stamford Bridge	42 449
Liverpool	Anfield	45 362
Manchester City	City of Manchester	48 000

   b *A possible solution*:

   Use Molineux, Goodison Park, Stamford Bridge.

   These can hold around the expected numbers or more and will have more atmosphere than Anfield or Manchester City as they will be closer to full.

   Other possible solutions:

   - Use grounds close by to save teams travelling.
   - Use grounds that are more spread out geographically to enable a wider spread of fans.
   - Use larger, more prestigious grounds, such as Manchester City.

5 14 043

6 **a** 5 m above sea level
   b 10 m below sea level

7 London and Southampton

8 **a** +20   **b** −20   **c** −5   **d** +5

9 Insufficient skirting board by 5 m

10 12 × £1350 = £16 200

11 3 days will be needed.

12 135 ÷ 28 = 4.8 so 5 coaches will be needed
   5 × 28 = 140 seats on 5 coaches
   140 − 135 = 5 spare seats

Chapter 2: Fractions, decimals and percentages

1 £5.13

2 17p more expensive

3 11 litres (or 12 litres as approximate only)

4 Jack's clothes store (£36)

5 Bella Belissima! (£17.25 compared with £20.50)

Chapter 3: Ratio and proportion

1 20:4 = 5:1

2 27:9 = 3:1
   This is not the same as 1:2. She is incorrect.

3 1200 ÷ 100 = 12 miles.

4 4.5 capfuls

5 Office cleaner: £6 per hour for 16 hours work. Total: £96

   Shop assistant: £5.50 per hour for 22 hours per work. Total: £121

   Advise Betty that the office cleaning job pays 50p more per hour and is only over 4 mornings per week. The shop assistant job is 6 hours more work and pays £25 more per week even though it is 50p less per hour. Which job she takes depends on whether how long she works for or how much money she earns is more important to her.

6 100 metres

Chapter 4: Time

1 3 hrs 25 mins

2 1hr 10 mins

3 He could start his holiday any day from 27th June – 1st July

4 Thursday

5 Tuesday 8th April

6 2105

7 1115

Chapter 5: Measures

1 1750 ml

2 340 g

3 12 cm

4 2.8°C

5 60.5 kg

6 4 to 6 litres

7 237 or 238 volts

Chapter 6: Drawing and measuring

1 4 cm

2 4.5 cm

3 Length 26.5 cm
   Width 20 cm (These measurements may vary slightly)

4 accurately drawn rectangle

Chapter 7: Formulae

1 2050

2 4 529 250 litres per day

3 Hours worked = 9
Number of cars sold = 2
Amount earned (£) = hours worked × 6
+ number of cars sold × 50
Amount earned (£) = 9 × 6 + 2 × 50
= 54 + 100
= £154

4 Number of days = 3
Number of extra hours = 4
Cost (£) = number of days × 10 + number of extra hours × 2 + 5
Cost = 3 × 10 + 4 × 2
Cost = 30 + 8
= £38

Chapter 8: Area and perimeter

1 14 m²

2 Approximately 6.5 km²

3 One stride = 80 cm
10 Strides = 80 × 10 = 800 cm = 8 m
18 strides = 80 × 18 = 1440 cm = 14.40 m
Area of classroom = 8 × 14.40 = 115.20 m²

Chapter 9: Collect and represent data

1 ☆ ☆ ☆ ☆

2 *A possible solution:*

Car colour	Tally	Frequency
Silver		
Black		
Red		
Blue		
Green		
Other		

3
Table number ....
Tea ☐
Coffee ☐
Continental ☐
English ☐
Breakfast ☐

4

Preferred chocolate brand	Tally	Frequency
A	ⅢⅢ Ⅲ	13
B	ⅢⅢ	10
Total	ⅢⅢ ⅢⅢ Ⅲ	23

5 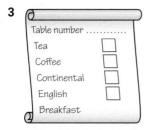 = 5 tourists

6 *A possible solution:*

Type of Vehicle	Tally	Frequency
Car		
Van		
Bus		
Taxi		
Other		
	Total	

Chapter 10: Use and interpret data

1 Range = 96 − 79 = 17p

2 Cost for adult: £238 × 2 = £476
Cost for children £125 × 2 × 2 = £500
Total cost for the family: £976

3 *A possible solution:*
10 calls in 40 minutes is an average of
4 minutes per call $\left(\frac{40}{10}\right)$. This is less than
5 minutes per call so she is meeting her target.

4 The average cost of a tyre is £52.75

5 *A possible solution:*
Mean amount taken = $\frac{£698.50}{3}$ = £232.83.
This is over £200 so Keith makes a profit.

Another possible solution:
Keith needs to take £200 a day for 3 days
= £600 to make a profit.
He takes £698.50. So he makes a profit.

6 *A possible solution:*
Assuming that Rosie wants the weather to be as hot as possible.
She should choose either Graffe or Miramar as the mean midday temperatures are both 34.1 (which is higher than Tinmouth)
Comparing just Graffe and Miramar, she should choose Miramar as the range in temperature is only 2°C rather than 12°C.

7 *A possible solution:*
Add up the totals for the week.
65 + 24 + 13 + 14 + 23 + 78 + 12 = 229 copies
(Take the opportunity to discuss with students why the graph is shaped like it is.)

Chapter 11: Likelihood of events

1 green **2** unlikely **3** unlikely
4 impossible certain

A: It is very unlikely that it will be light all night in England.
D: When you throw a dice, it could land on 1, 2, 3, 4, 5 or 6.
B: A coin is equally likely to land on heads or tails.
C: If you live a healthy life, it is likely you will reach the age of 60.